PENGUIN BOOKS

THE ELECTRONIC REPUBLIC

Lawrence K. Grossman is the former president of the
Public Broadcasting Service and of NBC News. He is
currently president of the Horizons Cable Network.
He lives in Westport, Connecticut, and works in New
York City.

324.63
Gro

DEC 1998

Farmers Branch Manske Library
13613 Webb Chapel
Farmers Branch, TX 75234-3756

LAWRENCE K. GROSSMAN

A TWENTIETH CENTURY FUND BOOK

PENGUIN BOOKS

THE
ELECTRONIC
REPUBLIC

RESHAPING DEMOCRACY IN THE

INFORMATION AGE

Farmers Branch Manske Library
13613 Webb Chapel
Farmers Branch, TX 75234-3756

TO ALBERTA N. GROSSMAN

PENGUIN BOOKS
Published by the Penguin Group
Penguin Books USA Inc., 375 Hudson Street, New York, New York 10014, U.S.A.
Penguin Books Ltd, 27 Wrights Lane, London W8 5TZ, England
Penguin Books Australia Ltd, Ringwood, Victoria, Australia
Penguin Books Canada Ltd, 10 Alcorn Avenue, Toronto, Ontario, Canada M4V 3B2
Penguin Books (N.Z.) Ltd, 182–190 Wairau Road, Auckland 10, New Zealand

Penguin Books Ltd, Registered Offices: Harmondsworth, Middlesex, England

First published in the United States of America by Viking Penguin,
a division of Penguin Books USA Inc. 1995
Published in Penguin Books 1996

1 3 5 7 9 10 8 6 4 2

Copyright © Lawrence K. Grossman, 1995
All rights reserved

A portion of the introduction first appeared in *Media Studies Journal*.

THE LIBRARY OF CONGRESS HAS CATALOGUED THE HARDCOVER AS FOLLOWS:
Grossman, Lawrence K.
The electronic republic: reshaping democracy in the information age
Lawrence K. Grossman.
p. cm.
Includes bibliographical references and index.
ISBN 0-670-86129-4 (hc.)
ISBN 0 14 02.4921 4 (pbk.)
1. Political participation—United States.   2. Democracy—United States.
3. Communication—Political aspects—United States.   4. Information society—
Political aspects—United States.   I. Title.
JK1764.G76   1995
324.6′3′0973—dc20   95–1449

Printed in the United States of America
Set in Sabon
Designed by Katy Riegel

Except in the United States of America, this book is sold subject to the condition
that it shall not, by way of trade or otherwise, be lent, re-sold, hired out, or otherwise
circulated without the publisher's prior consent in any form of binding or cover other
than that in which it is published and without a similar condition including this
condition being imposed on the subsequent purchaser.

ACKNOWLEDGMENTS

Most acknowledgment pages that I have read end where mine fittingly begins. I offer a special tribute to the clearheadedness, persistence, and dedication of my wife, Alberta, devoted teacher, demanding critic, and untiring supporter.

This book got its start after I left NBC News in 1988, when my longtime friend and colleague Marvin Kalb, director of the newly formed Joan Shorenstein Barone Center on the Press, Politics and Public Policy at the Kennedy School of Government, offered me temporary refuge in Cambridge in the Frank Stanton First Amendment Chair. Marvin made it a condition of my occupancy not only that I develop and teach a course but also that I write a paper, probably to make up for my absence of appropriate academic credentials. I called the course "Television and the Limits of the First Amendment," and I have no doubt I learned at least as much from my Kennedy School students as they learned from me, which made the course a success at least as far as I was concerned. The paper, "Reflections on Television's Role in American Presidential Elections," was published by the Kennedy School in January 1990.

After my stint as visitor at Harvard, I was invited to return to my own alma mater, Columbia University, as senior fellow at the Gannett Center for Media Studies, now called the Freedom Forum Media Studies Center. Under the remarkably civilized and supportive direction of

Everette E. Dennis, the center afforded me the opportunity to think through many of the issues raised in the succeeding pages. In addition to the lively, provocative company of my fellow fellows at the Gannett Center, I am especially indebted to its congenial and able staff, particularly Lisa DeLisle, Wendy Boyd, and Deborah Rogers.

John H. McMillan, retired Gannett editor and publisher who was also a senior fellow at the center, virtually adopted me and my project from the very first day and was indefatigable in reading draft after draft, making suggestions, raising questions, challenging assumptions, improving language, and insisting that the job get done. He is a remarkable human being. Other "senior" colleagues at the center, Leo Bogart, Donald M. Gillmor, and Gerald S. Lesser, were also of great help and encouragement.

The Twentieth Century Fund, co-publisher, has offered immeasurable editorial assistance, with persistent, intelligent, and, fortunately, often skeptical suggestions from its enormously supportive president, Richard C. Leone, and his colleagues. Members of the Fund's staff Jon Shure and Michelle Miller-Adams raised questions and poked holes in the drafts as they came in. Former Fund chairman Brewster C. Denny, Professor and Dean Emeritus at the Graduate School of Public Affairs of the University of Washington, read a very early version and gave his scholarly criticism and encouragement.

My colleague from public television Ed Pfister, the former president of the Corporation for Public Broadcasting and now Dean of the School of Communications at the University of Miami, generously arranged a campus seminar on electronic democracy for me during my week as visiting scholar there, long before the subject had much currency.

Friends and colleagues who were especially generous with their time and willingness to read and comment on the manuscript were Henry Geller, Professor Emeritus Henry F. Graff (my undergraduate Columbia College adviser, believe it or not), Ancil H. Payne, Richard C. Wald, and Tom Wolzien. I shamelessly have taken suggestions, ideas, and information from them all.

Others who have given their time for interviews, discussions, responses to my questions, requests for information, and general encouragement include: Diane Asadorian, Sandy Baron, Les Brown, Edith C. Bjornson, R. O. Blechman, Bernice Buresh, Robert Corn-Revere, Charles Firestone, Ed Fouhy, Beverly Goldberg, Richard Green, my brothers Dan and Richard Grossman, Ellen Hume, Brian Kahin, Mitch Kapur, Bill Kovach, Craig LaMay, Emanuel Margolis,

Gale Metzger, Newton N. Minow, Russ Neuman, Leona Nevler, Everett Parker, Tom Patterson, Monroe Price, Jason Renker, and Tracy Westin.

My thanks, too, to my agent, Jane Gelfman, who was absolutely determined to see that this book got into print; my editor at Viking, Marion Maneker, a man of considerable patience, enthusiasm, and supportiveness; Viking's publisher, Barbara Grossman (no relation), who thought up the title; and my talented colleague from CBS days Lou Dorfsman.

Finally, this page would not be complete without a special nod to the late Erwin Glikes, who instantly and enthusiastically took an intense interest in this book, nurtured its progress, and insisted that its author keep typing and get it done.

CONTENTS

FOREWORD

America's history reflects a deep mistrust of authority. The strong, representative national government championed by the framers of the Constitution was a hard sell. Over the years, the Founding Fathers (helped along by generations of historians in search of "truth" and politicians in search of legitimacy) have been interpreted as disagreeing sharply and fundamentally about just what should be the right mix of Jeffersonian Democracy and Hamiltonian Republic. But in their own time, Franklin, Hamilton, Madison, Jay, and others put aside their differences to fight for the Constitution. They were convinced that a federal republic was preferable to more direct and decentralized versions of democracy.

The debate over our system of governance has never ended. A century ago, William Jennings Bryan and others, feeding off the discontent caused by new immigration, industrialization, and, especially, the decline of the family farm, revived the founders' debate in new clothes: populism. Today, in even more modern dress, a form of populism is back. Not surprisingly, wrenching changes in the economy and the larger culture have spawned a strong reaction, which includes a revival of the belief that government is too powerful and inaccessible for the average citizen.

The New Populism, like the old, wants to alter the role of elected officials and enhance direct democracy at the expense of the representative process. More the product of nostalgia for a half-imagined past than of visions of a utopian future, the populist surge is evident in the popularity of new ballot initiatives, referenda, recall movements, and other expressions of direct democracy. A related phenomenon is the growing support for measures that constrain legislatures and the legislative process through term limits, balanced budget amendments, super-majority voting, and curtailment of staff support.

But what is truly different about today's populism is that technology is making mass participation possible in previously unimaginable ways. The electronic town hall, instant public-opinion polls, citizen juries, and other breakthroughs based, in part, on new communications technologies are touted as solutions to bad government and public cynicism. We seem to be in the early stages of a major shift away from our republican form of government. If this is the case, it's important to ask if direct "electronic" democracy is either workable or desirable in a country like America.

In this book, Lawrence K. Grossman provides his special insight into the current sweeping changes in media and politics, significantly expanding the essential public conversation about their implications. While acknowledging their deep disenchantment with the legislative process and politics in general, he insists that his readers take the next step and confront the most important implications of enlarged direct democracy: the extent to which it would be meaningless—even dangerous—without a concerned and educated public.

The very essence of democratic education and practice in this country is transmitted through an intermediating sea of news, information, and, increasingly, "infotainment." These complex and omnipresent media offer much of what shapes our perceptions about the nation and the world. Media are undergoing a transformation through the growth of the Internet and other computer links, as well as the beginnings of true interactivity in transmitting and transmuting information. Whether we will be enriched by this expansion of services is not at all clear; for it is entirely possible that, as with television, the flashy new computer screens will simply "occupy more of our time."

In the pages that follow, Grossman asks us to face some of the far-reaching consequences that are likely to accompany the spread of electronic connections in our society. That these changes are already having an effect is undeniable. That they are sure to continue is inevitable because of two characteristics of our system that we hold dear: individual liberty and free enterprise. The real question, then, may be not whether an increase in electronic democracy is desirable, but rather how to make the best of what is sure to be a part of our future.

Grossman also explores critical elements of the public culture of the United States—a culture that is rapidly changing and, some would say, deteriorating into a world of skepticism and scandal. He by no means "blames" media for sensationalizing our politics; after all, media do not set the national social agenda, nor do they tell politicians how to run their campaigns. Still, he does not excuse media professionals from some responsibility for the changes in our political culture.

For Grossman this book ties together the work of a lifetime. He has served in senior posts in the private and public sectors, notably as president of NBC News and as president of PBS. Significantly, however, what he brings to this effort is not only his professional expertise but his deep commitment as a citizen. He offers a serious agenda for communications reform, including specific recommendations that, in effect, challenge government and business—but most of all, citizens in general—to take control of shaping this electronic future.

Grossman also is candid and forceful in recommending reforms in the news business, especially in its dominant form, television. He makes a strong case, for example, for offsetting the notoriously short attention span of commercial television news with expanded publicly supported coverage of public issues. His message is clear: the most democratic and responsive country in the world is about to undergo a vast shift toward more direct participation—at least by those who choose to participate. But unless citizens are educated about public issues, unless they think beyond their individual and sometimes selfish concerns, we may be headed for a new kind of trouble: a tyranny of whim that not only the Founding Fathers but even more recent leaders could not have foreseen.

Finally, history and democratic theory tell us that an informed and interested public is the key to successful self-governance. Larry Grossman's book provides a powerful reaffirmation that, for all the potential that modern technology provides for fulfilling the dream of direct democracy, the necessity for an engaged and knowledgeable citizenry remains the foundation of all our hopes. On behalf of the Trustees of the Twentieth Century Fund, I thank him for his efforts.

Richard C. Leone
President
The Twentieth Century Fund
January 1995

The Twentieth Century Fund sponsors and supervises timely analyses of economic policy, foreign affairs, and domestic political issues. Not-for-profit and nonpartisan, the Fund was founded in 1919 and endowed by Edward A. Filene.

BOARD OF TRUSTEES OF THE TWENTIETH CENTURY FUND

Morris B. Abram, *Emeritus*

H. Brandt Ayers

Peter A. A. Berle

José A. Cabranes

Joseph A. Califano, Jr.

Alexander Morgan Capron

Hodding Carter III

Edward E. David, Jr.

Brewster C. Denny, *Emeritus*

Charles V. Hamilton

August Heckscher, *Emeritus*

Matina S. Horner

Lewis B. Kaden

James A. Leach

Richard C. Leone, *ex officio*

P. Michael Pitfield

Don K. Price, *Emeritus*

Richard Ravitch

Arthur M. Schlesinger, Jr., *Emeritus*

Harvey I. Sloane, M.D.

Theodore C. Sorensen, *Chairman*

James Tobin, *Emeritus*

David B. Truman, *Emeritus*

Shirley Williams

William Julius Wilson

Richard C. Leone, *President*

PART ONE

INTRODUCTION

A new political system is taking shape in the United States. As we approach the twenty-first century, America is turning into an electronic republic, a democratic system that is vastly increasing the people's day-to-day influence on the decisions of state. New elements of direct democracy are being grafted on to our traditional representative form of government, transforming the nature of the political process and calling into question some of the fundamental assumptions about political life that have existed since the nation was formed more than two hundred years ago.

The irony is that while Americans feel increasingly powerless, cynical, and frustrated about government, the distance between the governed and those who govern is actually shrinking dramatically. Many more citizens are gaining a greater voice in the making of public policy than at any time since the direct democracy of the ancient Greek city-states some twenty-five hundred years ago. Populist measures such as term limits, balanced budget amendments, direct state primaries and caucuses, and expanding use of ballot initiatives and referenda reduce the discretion of elected officials, enable voters to pick their own presidential nominees, bypass legislatures, and even empower the people to make their own laws. Incessant public-opinion polling and increasingly sophisticated interactive telecommunications devices make government instantly aware of, and responsive to, popular will—some say, too responsive for the good of the nation. As the elect seek to respond to every twist and turn of the electorate's mood, the people at large are taking on a more direct role in government than the Founders ever intended.

This democratic political transformation is being propelled largely by two developments—the two-hundred-year-long march toward political equality for all citizens and the explosive growth of new telecommunications media, the remarkable convergence of television, telephone, satellites, cable, and personal computers. This is the first generation of citizens who can see, hear, and judge their own political leaders simultaneously and instantaneously. It is also the first generation of political leaders who can address the entire population and receive instant feedback about what the people think and want. Interactive telecommunications increasingly give ordinary citizens immediate access to the major political decisions that affect their lives and property.

The emerging electronic republic will be a political hybrid. Citizens not only will be able to select those who govern them, as they always have, but increasingly they also will be able to participate directly in making the laws and policies by which they are governed. Through the use of increasingly sophisticated two-way digital broadband telecommunications networks, members of the public are gaining a seat of their own at the table of political power. Even as the public's impatience with government rises, the inexorable progress of democratization, together with remarkable advances in interactive telecommunications, are turning the people themselves into the new fourth branch of government. In the electronic republic, it will no longer be the press but the public that functions as the nation's powerful "fourth estate," alongside the executive, the legislative, and the judiciary.

The rise of the electronic republic, with its perhaps inevitable tendency to respond quickly to every ripple of public opinion, will undercut—if not fundamentally alter—some of our most cherished Constitutional protections against the potential excesses of majority impulses. These protections were put in place by the Founders, who were as wary of pure democracy as they were fearful of governmental authority. The Constitution sought not only to protect the people against the overreaching power of government but also to protect the new nation against the overreaching demands of ordinary people, especially the poor.

Telecommunications technology has reduced the traditional barriers of time and distance. In the same way it can also reduce the traditional

Constitutional barriers of checks and balances and separation of powers, which James Madison thought the very size and complexity of the new nation would help to preserve. "Extend the sphere, and you take in a greater variety of parties and interests; you make it less probable that a majority of the whole will have a common motive to invade the rights of other citizens." However, as distances disappear and telecommunications shrink the sphere, and as the executive and legislative branches of government become more entwined with public opinion and popular demand, only the courts may be left to stand as an effective bastion against the tyranny of the majority. The judiciary, the branch of government that was designed to be the least responsive to popular passion, will bear an increasingly difficult and heavy burden to protect individual rights against popular assault.

Direct democracy, toward which we seem to be inexorably heading, was the earliest form of democracy, originating during the fifth century B.C. in the small, self-contained city-states of classical Greece. During the two hundred years of Athenian direct democracy, the ancient city-state whose governance we know most about, a privileged few citizens served at one and the same time as both the rulers and the ruled, making and administering all their own laws. "Although limited to adult males of native parentage, Athenian citizenship granted full and active participation in every decision of the state without regard to wealth or class." Democracy in Athens was carried as far as it would go until modern times.

By contrast, representative government—democracy's second transformation—is a relatively recent phenomenon, originating in the United States a little more than two centuries ago. Under representative democracy, Americans—at first a privileged few and now every citizen over age eighteen—can vote for those who make the laws that govern them. Unlike the ancient Greeks, our Constitution specifies a government that separates the rulers from the ruled. It connects the people to the government by elections, but distances the government from the people by making the elected the ones who actually enact the laws and conduct the business of government. As one political scientist put it, that "Constitutional space is the genius of American republicanism. It keeps the process of democratization under control and prevents our democracy from ruining itself by carrying itself to extreme."

Today, that constitutional space is shrinking. New populist processes and telecommunications technologies amplify the voice of the people at large and bring the public right into the middle of the decision making processes of government. As the power of public opinion rises, the roles of the traditional political intermediaries—the parties, the mass media experts, and the governing elite—decline. Institutions that obstruct the popular will or stand between it and the actions of government get bypassed.

Telecommunications technologies—computers, satellites, interactive television, telephone, and radio—are breaking down the age-old barriers of time and distance that originally precluded the nation's people from voting directly for the laws and policies that govern them. The general belief holds that representative government is the only form of democracy that is feasible in today's sprawling, heterogeneous nation-states. However, interactive telecommunications now make it possible for tens of millions of widely dispersed citizens to receive the information they need to carry out the business of government themselves, gain admission to the political realm, and retrieve at least some of the power over their own lives and goods that many believe their elected leaders are squandering.

The electronic republic, therefore, has already started to redefine the traditional roles of citizenship and political leadership. Today, it is at least as important to reach out to the electorate—the public at large—and lobby public opinion, as it is to lobby the elect—the public officials who make the laws and administer the policies. In the words of literary critic Sven Birkerts in *The Gutenberg Elegies*, "The advent of the computer and the astonishing sophistication achieved by our electronic communications media have together turned a range of isolated changes into something systematic. The way that people experience the world has altered more in the last fifty years than in the many centuries preceding ours." The emergence of the electronic republic gives rise to the need for new thinking, new procedures, new policies, and even new political institutions to ensure that in the century ahead majoritarian impulses will not come at the expense of the rights of individuals and unpopular minorities.

We need to recognize the remarkable change that the interactive telecommunications age is producing in our political system. We need to understand the consequences of the march toward democratization. We need to deal with the promise and perils of the electronic republic. It can make government intensely responsive to the people. It also can carry responsiveness to an extreme, opening the way for manipulation, demagoguery or tyranny of the majority that "kindle[s] a flame . . . [and] spread[s] a general conflagration through the . . . States."

Most studies of government, politics, and the media start at the top by examining the qualities of leadership that define political life. The message of this book, however, is that in the coming era, the qualities of citizenship will be at least as important as those of political leadership. In an electronic republic, it will be essential to look at politics from the bottom up as well as from the top down.

This book will explore both the opportunities and the dangers ahead for participatory democracy in the electronic information age. It will suggest ways to organize our political system to embrace the new forms of government without undermining or destroying this country's essential constitutional protections and values.

What will it take to turn the United States into a nation of qualified citizens who are engaged not as isolated individuals pursuing their own ends but as public-spirited members who are dedicated to the common good? In an electronic republic, finding the answer to that question is essential. In the words of Thomas Jefferson, "I know of no safe depository of the ultimate powers of the society but the people themselves, and if we think them not enlightened enough to exercise their control with a wholesome discretion, the remedy is not to take it from them, but to inform their discretion."

In the spirit of Jefferson, this book is designed to help inform the people's discretion. Part I examines representative democracy and the end of the old politics. Part II considers the electronic republic that has already begun and offers suggestions to help deal with the inevitable changes it will bring.

1

TRANSFORMING DEMOCRACY— AN OVERVIEW

In January 1991, I spent the first day of America's entry into the Persian Gulf War in a remote Swiss army reserve training camp on the outskirts of the tiny skiing village of Andermatt, deep inside the Swiss Alps. I had been invited there to talk to a corps of Swiss army information officers about the role of television in time of war. In civilian life, these were Switzerland's top newspaper editors and broadcast executives, an obviously prosperous, influential, and well-fed group of reserve military officers whose members did not fit my image of the lean, well-conditioned Swiss citizen army. They planned to spend this training period discussing the fast-changing international world of telecommunications and the complex relationship between television, the government, and the military. And they exhibited a professional curiosity about my own experience in the United States because I had recently spent time in both government-supported public television, as president of PBS, and in private commercial television, as president of NBC News.

An actual world crisis—the entry of the United Nations into the war in the Persian Gulf on my first day there—gave a sense of urgency and instant reality to our discussion. The Swiss army information officers and I spent the seminar's first hours tuned to CNN's dramatic coverage of the sudden widening of the war, with live reports from Washington, the Persian Gulf, Israel, Moscow, and the capitals of Europe. Although I had spent the entire previous decade and more trying to keep up with the revolutionary changes being wrought by new telecommunications technology and orchestrating PBS's and then NBC News's responses to those changes, my experience as an ordinary television viewer in this hidden and remote army camp high in the Swiss Alps was both exhilarating and terribly frustrating.

Via special satellite feed, we could see government leaders, military personnel, and people on both sides and throughout the world reacting to the international crisis. We had a sense that we ourselves were actually participating in the events in some way, by watching them on the glass screen as they took place. Yet we could see nothing of the Gulf War itself on television. That morning the world had indeed become a global information village as we all shared direct access to the historic adventure taking place thousands of miles away. But the global village was also a global rumor mill, as television's reporters, commentators, anchors, and experts filled hours of live airtime speculating about distant events in the Gulf that they could not see and knew hardly anything about. Finally, it was also an exercise in global manipulation as everyone on the screen—reporters, citizens, experts, and government officials—closed ranks to support the troops going off to war.

In March 1991, barely a week after America's first live, in-color television war ended in a lightning-quick victory over Saddam Hussein's Iraq, I traveled to Pittsburgh to deliver the annual American Experience lecture at the University of Pittsburgh, a major event. In my Pittsburgh talk, I had planned to reflect on my unlikely experience of watching the war in the Persian Gulf against Iraq unfold in the army camp in the Swiss Alps and, on the basis of that, then look ahead to what might be expected from television's coverage of the 1992 presidential election campaign, which was just about to begin. Television's role in reporting the war and its role in reporting presidential election politics seemed to me at the time to be not as far apart as one might expect. While the remarkably vivid live coverage of the Gulf War was still uppermost in many people's minds, much of it—as we had already learned by then—was substantially wrong, misleading, and manipulated. Similarly, television's abysmal coverage of the 1988 presidential election also had opened major questions, still being widely debated at that time, about the integrity of the process and the direction of television news. I planned to focus on those two widely disparate events in my American Experience lecture in Pittsburgh in order to raise the question whether television, for all its technological advances, was getting better or worse as it moved "Beyond the Wasteland."

As it happened, the university's genial president, Dr. Wesley W. Posvar, stole my thunder. I met Dr. Posvar at a dinner reception immediately before my lecture. He had been a career military officer prior to becoming the university's president and our conversation zeroed in immediately on television's coverage of the Gulf War. Dr. Posvar, like everyone else, had been mesmerized by television's remarkable live images of the Scud and Patriot missiles dueling in the skies high above the Persian Gulf.

The two of us compared notes on how much television technology had changed since the Vietnam War, and even in the almost three years since I had left the presidency of NBC News. In the Gulf War, unlike in Vietnam, the American people saw a great deal of live coverage but none of the carnage and devastation on the ground. The vivid satellite pictures from the Persian Gulf looked more like images from *Star Wars* or a Nintendo video game than real life. Our conversation prompted Dr. Posvar, in introducing me to the Pittsburgh audience, to speculate on whether future wars, like current presidential election campaigns, might actually be fought entirely on television screens instead of on battlefields. "Cyber-warfare," he suggested, could "use the ultimate 3-D techniques of virtual reality, electronic simulation, and interactive video. Televised war would look, sound, and eventually perhaps even feel and smell so much like the real thing that real war would no longer be necessary." In the same spirit, I offered a scenario for the start of the next war: Every television screen in the nation will switch on automatically with a special crisis alert. A phone number, "1-800-U-DECIDE," would flash on and off while the President, flanked by the secretary of defense, chairman of the Joint Chiefs of Staff, and chiefs of all the armed services, would appear simultaneously on every channel, preempting every show, even *Oprah*. Using maps, charts, and electric pointers, they would brief the nation's citizens on the crisis. Then the president would end by asking all citizens to help their commander in chief and the Congress arrive at the right decision: "Press 1 if you think we should go to war. Press 2 if you think we should stay out of it. Press 3 if you want to tell us which targets to hit. Call 1-800-U-DECIDE, the number flashing below, for more information or to discuss any questions you may have."

A year before the arrival of H. Ross Perot on the campaign trail, Dr. Posvar anticipated a scenario for the emerging American electronic republic. We have already reached the point where no major political decision gets made without first taking the temperature of the American public. It will not be long before many Americans sitting at home or at work will be able to use telecomputer terminals, microprocessors, and computer-driven keypads to push the buttons that will tell their government what should be done about any important matter of state.

THE RISE OF THE ELECTRONIC REPUBLIC

Until very recently, George Orwell's nightmare tale of Big Brother who utilized electronic surveillance technologies to monitor every citizen, hear every word being said, and see everything being done ("Big Brother Is Watching You!") was the prevailing metaphor for the century to come. The frightening vision of Orwell's 1984 evaporated with the disintegration of the monolithic Nazi and Communist regimes.

By contrast, the twenty-first century's defining image is more likely to have ordinary citizens using their personal telecommunications devices to keep Big Brother under continuing surveillance. The tables have turned. For the nation's leaders, as First Lady Hillary Clinton complained, "There is hardly any zone of privacy left anymore." With the push of a teleprocessor button or the stroke of a telecomputer key, citizens can tell their leaders exactly what they think and what they want them to do. In the "smart" media world, information no longer flows only one-way from one to many. Instead, it flows simultaneously and instantaneously in many directions, from the bottom up as well as from the top down.

For many of my well-educated colleagues in the media and public affairs, and, it seems, for most political scientists, this second scenario is not much more reassuring than the first. Though they profess to be enthusiastic supporters of democratic governance, they actually fear the prospect of too much democracy as much as they fear too little. Like most of our nation's Founders, they do not trust the unmediated judgment of the people at large. They prefer "to refine and enlarge the public views, by passing them through the medium of a chosen body of citizens, whose wisdom may best discern the true interest of their country, and whose patriotism and love of justice will be least likely to sacrifice it to temporary and partial considerations." The world is an increasingly complicated and dangerous place, they say. Problems that require solutions are fast-moving and complex, and growing more so every day. To make intelligent decisions about the multidimensional issues that face us, we need experts with specific knowledge, sophistication, experience, and appropriate educational background. Ordinary people do not have the time or inclination to delve into the details of health care reform and crime legislation or become informed about our trade imbalance or Middle East diplomacy. As much as one would like to believe that the people themselves are best qualified to judge what is in their self-interest, the reality is that informed specialists are more likely to make sound judgments. Democracy needs a governing elite. Of course, the people at large tend to have a very different view these days about the "wisdom" and "love of justice" of their "chosen body of citizens."

Twenty-five hundred years ago Plato, drawing on his own experiences in Athens, expressed similar distrust of the common people in *The Republic.* The people, he said, are bad judges in political matters. The common man has no experience with foreign policy, economics, or national issues; to expect sensible judgment and expert knowledge on such matters from ordinary citizens "is to expect the impossible. He will judge on impulse, sentiment, or prejudice, and though his heart may be sound his head will be muddled." Moreover, direct democracy, Plato said, encourages bad leadership. People cannot be trusted to make the best choice of leaders, and since the popular leader depends entirely on popular favor, he "will constantly be tempted to retain that favor by the easiest possible means." How? "He will play on the likes and dislikes, the weaknesses and foibles of the public, will never tell them an unpleasant truth or advocate a policy that might make them uncomfortable. . . . The people care little who their leaders are provided they profess themselves 'the people's friends.' " Thus, "popular leaders are as devoid of true knowledge as are the people they lead." What is worse, according to Plato, in a pure democracy, where "every individual is free to do as he likes . . . , the minds of the citizens become so sensitive that the least vestige of restraint is resented as intolerable." Plato may as well have been describing today's efforts by democratic leaders to placate a restless and dissatisfied public in which "fathers pander to their sons, teachers to their pupils." Therefore, "the public voice pronounced by the representatives of the people, will be more consonant to the public good, than if pronounced by the people themselves." So goes the classical argument against pure democracy.

If the new telecommunications age brings unmediated democracy, what will happen to our carefully contrived constitutional system of checks and balances? Who will protect minorities against the passions and tyranny of the popular majority? Who will protect the majority from manipulation by public opinion experts, political spin doctors, and unscrupulous pollsters? Who will protect the poor from the permanent majority of the "haves?" Who can offset the persuasive power of big money, the often lying and misleading political commercials, the corruption of politics by what *New York Times* columnist Russell Baker and others have called the "legal bribery" of lobbyists and political action committees seeking favors from government, and the pervasive influence of those who control big media? In a pure democracy, how can those who own the means of communications or dominate the media by lopsided expenditures be prevented from overwhelming the debate, slanting the discussion, and unfairly influencing the public at large?

What chance do ordinary citizens have to come to sound public judgments on health care reform, for example, when, given the unlimited sums of money available for lobbying and political campaigns by those with large-scale financial professional interests, even "the Oval Office . . . [was] reduced to just another trade association?"

Opponents of direct democracy cite survey after survey revealing the depth of people's ignorance about public affairs: They cannot name their own state's U.S. senators, nor can they identify the chief justice of the U.S. Supreme Court, or the nation's principal cabinet officers. One survey found that if the Bill of Rights were on the ballot today, most people would vote against it. Political scientists, not without reason, disdain the quality and depth of news and information that people receive from television, their chief source of information about government and world affairs. Critics point to the consistently low ratings for the few remaining serious television documentary programs about important public issues, which invariably rank at the very bottom of television's popularity charts. And in the current commercial television climate such documentaries have all but disappeared from the screen. The quality of information is said to be getting worse, not better. These are not encouraging recommendations for expanding the role of public opinion in major decision making about affairs of state.

As another sign of the dangerous lack of public interest and involvement in politics, direct democracy's opponents recite the abysmally low turnout at elections. A bare majority of those eligible bothers to vote in presidential elections. A paltry minority turns out during bi-election years. In 1994, a record 39 percent of the voters went to the polls but that means more than 60 percent of the electorate did not even bother. Even if many of these dropouts were to be brought back into the political fold, the skeptics ask, are these the people we should entrust with the ultimate power to make public policy and decide our nation's fate—a mass of citizens of unequal and most often inadequate knowledge, training, wisdom, and interest? For most issues, it is unlikely they will take the time to learn all they need to know or deepen their understanding, no matter how inexpensive, accessible, or convenient additional information might be.

By contrast, those who advocate moving closer to "pure democracy," in which all the people govern themselves in public matters all the time, argue that the very process of representation serves to discourage participatory citizenship. By adapting pure democracy to the realities of governing large-scale nation states, representation gains efficiency without sacrificing accountability. But leaving the major decisions on public matters to a few elected representatives leaves the people out of direct decision making and diminishes public participation and involvement.

Nevertheless, stimulated by new and old interactive communications technologies and rising disgust over the way professional politicians, government employees, and elected incumbents function, many members of the public are seeking to gain greater control over the workings of government. They want to reduce the scope of politicians' responsibility through such measures as term limits, balanced-budget amendments, and state ballot initiatives. On the federal level, one-term administrations and split party government have become the rule, not the exception. On the state level, major policy decisions increasingly are being made in the voting booth, bypassing legislatures and governors in California, Oregon, Colorado, Massachusetts, and a growing number of states. Elements of direct democracy are being thrust into the political system. The major political institutions—the presidency, the Congress, state and local governments, the parties, the press, the process for nominating and electing political leaders—have to accommodate to this far-reaching new populist political reality. Even the judiciary is not immune from increasing public oversight and popular attention as courtrooms have been opened to the prying lenses of television cameras.

THE PUSH OF TECHNOLOGY

Each year millions more citizens acquire personal computers, go online with computer networks, subscribe to cable and satellite services, and use telephones and faxes to communicate their views on hot political issues to local, state, and federal officials. The spread of computers is changing the political process as radically as it is changing so many other aspects of our lives. Subscribers to CompuServe have a network that puts them on-line with the White House, which is also available to everyone with access to Internet. Members of Congress and state legislators opened computer mailboxes and send messages to constituents through computer bulletin boards and E-mail exchanges. In 1994, Minnesota candidates for governor and the U.S. Senate squared off in the first election campaign debates held by computer, giving voters more direct access than on the old-fashioned campaign trail. "The Internet and other spurs of the information superhighway have emerged as powerful new links between politicians and voters." Politics has invaded cyberspace. A new breed of professional grassroots lobbying firms, using computerized lists and telephone banks in "boiler rooms," has geared up to mobilize the pressure of public opinion at the behest of paying clients. Government officials were always fair game to be bought by special interests. Now that public sentiment plays a central role in the making of major political decisions, it too is being bought and delivered—for a price—to members of Congress and the White House.

But computers are only one part of the story of the revolution in communications. Talk radio and talk television hosts, mostly but by no means all conservative, are proliferating and gaining influence and popularity. Today, most have added old-fashioned telephone lines to their broadcasts, opening the airwaves to members of the audience at home and provoking listeners and viewers to bombard Congress with grassroots opinions on controversial issues—congressional pay raises, abortion, welfare policy, Social Security, immigration, tax policy, and health care, for example. Developing public feedback has turned into a major political and media industry.

The big losers in the present-day reshuffling and resurgence of public influence are the traditional institutions that have served as the main intermediaries between the government and its citizens—the political parties, labor unions, civic associations, even the commentators and correspondents in the mainstream press. Not only are they declining in influence and relevance, but surveys demonstrate that they are losing the public's trust. Sitting before their glass screens at home, people see for themselves what is happening in the world. They judge for themselves the quality and character of the political leaders who parade before them on television seeking their support. Who needs party bosses, union leaders, political columnists, or TV commentators anymore to tell people what to think and which way to vote? Today, the people stay home and make up their own minds while watching television. And the journalists spend a good deal of their time analyzing and reporting what the people think and how they feel.

To its critics, teledemocracy conjures the image of alienated, silent voters, sitting alone inside an electronic cocoon. Like television directors in a sealed TV control room, these modern-day robotic voters absorb virtually all their information about the outside world electronically—through CNN-style live coverage, tabloid-news shows, and sensational newsmagazines; and, increasingly, at programmed computer terminals. The wired public then feeds back its ill-formed, unsophisticated, unmediated opinion instantaneously, without deliberation, following simple on-screen instructions to press Y for yes, or N for no. These isolated electronic citizens never participate in the political community. They rarely discuss issues with fellow citizens face-to-face. Historian Daniel Boorstin contrasts the private, "segregated" experience of watching television, in which each person's experience is separate and silent, with the necessary healthy and vigorous deliberative "public life of a democracy." But is that actually the way electronic democracy will operate, or needs to operate in the future?

THE PEROT MODEL

Television's coverage already makes most people firsthand witnesses to wars, disasters, crimes, crises, and election campaigns. It gives viewers the sense of being direct participants in these events. And based on what these direct participants see with their own eyes and hear with their own ears, they prefer to come to their own conclusions about what should be done and which candidates to support, with little help, thank you, from political parties, journalists, experts, or most other authority figures.

A recent case in point is the rise and fall of multimillionaire H. Ross Perot, the arrogant and colorful on-again, off-again political neophyte from Texas who did the unthinkable. He started all by himself, ran for president as an independent without the support of a single identifiable political constituency, and was taken seriously enough to change the dynamics of the entire election campaign. A new breed of participatory media—CNN's *Larry King Live!,* local and network call-in television shows, talk radio, and televised town meetings—helped trigger Perot's meteoric candidacy. He let the genie of the new politics out of the bottle, shrewdly finding new ways to take advantage of these electronic media to reach out directly to the people without, in the beginning at least, letting any of the "professionals" get in the way. Perot's highly publicized populist ideas to govern by means of national plebiscites, televised town meetings, and public feedback—although vague, ill-formed, and possibly even demagogic—may well turn out to be his most important and enduring legacy. Media critics and political scientists universally derided Perot's notion of an electronic town hall. But it will be a reality in some form in the century ahead.

Appealing directly to voters by frequent personal appearances on syndicated talk shows, telephone call-ins, and "infomercials," Perot's low-tech, back-channel television campaign at first bypassed the traditional political intermediaries—the major parties and the mainstream press. Without spending a dime of his own money buying media or holding a single press conference in the initial phase of his campaign, Perot climbed to first place in the national polls only four months after announcing his availability on CNN's *Larry King Live!* —a television program with a minuscule audience compared to the nightly network newscasts and prime-time magazines. If Perot had run in the California primary in June, statewide polls suggested he could have won on either the Democratic or Republican ticket. Perot failed to maintain that momentum in part because persistent hard questioning and investigative reporting by the establishment press ultimately helped do him in, in part because his flip answers to hard questions wore thin, and in part because his colorful, authentically original personality lost its appeal with intense exposure over time. But Perot's trajectory of political popularity was still remarkable. It is an augury of the politics of fast change in the future.

While Perot's campaign style was new, his populist appeal was not. The Perot campaign was part of a strong populist tradition in American politics whose roots go back to the Revolutionary War. Perot's messages recall the *Common Sense* of Tom Paine, the populist rhetoric of Patrick Henry and later, the presidential campaign of Andrew Jackson, who promised to open the doors of the White House to ordinary people. In political style, if not content, Perot followed in the footsteps of Wisconsin Senator Robert M. La Follette, Jr., Socialist Eugene Debs, Louisiana's maverick Huey Long, and Alabama Governor George Wallace. "Go back to the first principles of democracy; go back to the people!" La Follette exhorted the students of Chicago University in 1897. Like La Follette and Debs, Perot also promised to "go back to the people" and make them part of the decision process. As president, Perot said, he would have the government do only what "the real folks out there" asked him to do, nothing more. He said he would capitalize on new interactive communications technologies to develop a people's consensus on major issues by conducting regularly scheduled electronic town meetings. He proposed that a national referendum should replace the Congress in voting on new federal taxes. And he even offered to resign, as president if enough citizens ever phoned and faxed the White House to say he should quit. In the end, Perot's populist appeal won him almost 20 percent of the national vote, more than any other independent presidential contender received since Theodore Roosevelt tried to regain the White House on the Bull Moose ticket. (Two years after Perot's presidential try, Italian Silvio Berlusconi, another political maverick with an overnight national political movement and extraordinary access to his nation's television screens succeeded where Perot had failed. Berlusconi, who was elected Italy's prime minister in 1994 under the banner of a brand new political party, had a great advantage over Perot: He owned both of Italy's private television networks, as well as major newspapers and a big advertising agency, and he used them all to the hilt to promote his political ambitions. Berlusconi's whirlwind political success is the best and most unnerving validation yet for A. J. Liebling's classic epigram, "Freedom of the press is guaranteed only to those who own one.")

THE PLEBISCITE PRESIDENCY:
THE CRITICS' VIEW

We shall look back on the 1992 presidential campaign and the live television coverage of the 1991 Gulf War as watershed moments— turning points in the decade-long rise of the American electronic republic. In the 1992 campaign, Perot's mainstream Democratic and Republican rivals for the White House followed his lead, tapping into many of the same direct-to-the-people campaign strategies. During the New Hampshire primary, the press had become so totally consumed with Bill Clinton's alleged affair with nightclub performer Gennifer Flowers that Clinton decided the only way to escape their "feeding frenzy" was to bypass the mainstream reporters and talk directly to the state's potential voters himself. He addressed the issues they were concerned with in a series of well-publicized town meetings on radio and television. Even President Bush eventually awakened to that strategy, when it became apparent that he was in trouble with the voters. Bush invited Larry King to originate his CNN call-in show directly from the White House late in the campaign.

In 1992, political commentators of every stripe, from liberal Anthony Lewis in *The New York Times* to conservative George Will in *Newsweek*, deplored this new populist electronic political trend. They were appalled at the idea that the United States should have a "plebiscite presidency" and "rule by applause meter." Direct electronic democracy, wrote columnist Charles Krauthammer, is the "highway to tyranny," signifying the end of reasoned, experienced, deliberative constitutional political decision making and sure to lead the country down the slippery slope to demagoguery.

These critics echoed the two-centuries-old arguments of Alexander Hamilton in the Federalist Papers. Difficult political decisions should not be left to the snap judgments and popular distemper of public opinion, Hamilton wrote. They should be made only by those who are the "most likely to possess the information and discernment requisite . . . to complicated investigations." In other words, major policy decisions should be left to an informed political elite, what historian James MacGregor Burns has called "the seasoned wisdom" of educated, informed, responsible leaders like the Founders themselves.

Notwithstanding these concerns, however, many of the constitutional roadblocks that originally were put in place to thwart instant responsiveness to majority rule have been pushed aside. The elitist republic has evolved into an inclusive democracy. During the past two centuries, American democracy has progressed inexorably toward greater political equality and universal suffrage. The right to vote, originally limited to a narrow group of adult white male property owners who comprised about 10 percent of the population, has been extended over two centuries to all citizens over age eighteen. The insulating role of the electoral college, originally designed to select presidents by a political elite rather than by popular vote, survives as a kind of quaint—although still potentially disruptive—vestigial political institution. U.S. senators, once chosen by state legislatures, have been elected by direct popular vote since 1913. Since the rise of television in the 1950s, and largely *because of* the rise of television, presidential nominees are no longer picked by party bosses in smoke-filled rooms or even by party delegates in national conventions. Instead, they are elected by rank-and-file voters in direct state primaries and caucuses.

As Tocqueville discovered, American democracy "is constantly in the process of democratizing itself further."

THE RISE OF BALLOT INITIATIVES

States and localities have gone even further than the federal government in democratizing decision making and making government more responsive to the people. In twenty-three states and the District of Columbia, ballot initiatives and referenda—a movement that started in California early in the century and has moved all across the country—bypass the long-standing lawmaking power of state and local legislatures. In these states, "voters now exercise many of the legislative and executive powers traditionally reserved for the first and second branches of government." In recent years, initiatives in California have given the state's entire electorate the opportunity to vote directly on issues as diverse as immigration policy, tax reduction, nuclear restraint, AIDS testing, school funding, insurance regulation, wildlife preservation, prison labor, the state lottery, election campaigning, state budget limits, and more. To some, the initiative and referenda have proliferated to the point that they already have become onerous and impractical. The initiative originated in South Dakota in 1898. In the 1994 election, Dr. Donald Balmer, chairman of the Political Science Department of Oregon's Lewis and Clark College, said that he cast fifty-one votes on his single ballot on initiatives, referenda, and candidates. To be expected to vote that many times and make that many decisions on complex issues and competing officials overwhelms even the most sophisticated voter. In the expanding roll call of states that use initiatives and referenda, no political issue is too big or too small to be decided by popular vote rather than by legislators, even though the latter presumably were elected to decide those issues on their constituents' behalf.

And their influence on national policy has been pronounced. California's famous Proposition 13, passed in 1978, limiting the state's taxing power, has since been adopted by numerous other states and profoundly influenced federal tax policy during the Reagan years. The same state's Proposition 187, voted in 1994, sharply restricting services to illegal aliens, similarly set the tone for the rest of the nation in re-examining federal immigration policy.

OPENING THE COURTS
AS WELL

Nor has the judiciary been immune from this populist trend that is reshaping the American democratic system. With television cameras now allowed in most courtrooms, judges and juries, once largely insulated from daily public scrutiny, must contend with millions of cable television viewers looking over their shoulders every day. At times, the television audience gets to see key criminal trial evidence on videotape long before it ever surfaces in the courtroom for the jury and judge to examine. In highly publicized cases, lawyers for both the defense and prosecution feel obliged to pursue tactics that will help convince not only the jury and judge in court but also the audience watching at home.

Not even *I Love Lucy* reruns received more television exposure than the grainy, graphic, black-and-white home video images of Los Angeles police officers crouched over motorist Rodney King, beating him with their batons. The officers' acquittal on all charges in their first trial shocked millions throughout the country who, even before the jury had been impaneled, watched the evidence on television being played and replayed in slow motion and decided that the accused were guilty as charged. The ensuing public outrage in response to the outcome of the first trial unquestionably led the U.S. Justice Department to bring new federal charges against the police, prompting a second trial that convicted two of the police officer defendants. Long before the O. J. Simpson murder trial got under way in Los Angeles, both sides consistently leaked evidence to the press, held news conferences, and did everything they could to turn television to their favor. The courts, too, have succumbed to the march of popular electronic scrutiny.

THE RISE
OF PUBLIC DISCONTENT

Politically, one problem has been that the more television viewers see of their government in action—or in gridlock—the less they like what they find. Whether most viewers are actually being shown more of their government at work, or whether they are being served only with more snippets of government failures and outrages is a question we shall discuss later on. But the fact remains that most Americans have grown increasingly disenchanted with politics, politicians, and political parties. Television has fed their cynicism. According to a Louis Harris poll, taken shortly before the 1992 presidential election, 72 percent of all voters believe their leaders are out of touch with the people, 66 percent feel a sense of powerlessness, and 61 percent say those who run the country don't care what happens to them.

That disillusionment, the deepening sense of alienation and frustration that citizens feel toward government, has served only to accelerate the demand for basic structural changes in the nation's political system. Some people tend to embrace whatever holds out the prospect of empowering them, while others simply withdraw from participation as citizens on the not unreasonable assumption that their involvement will make no difference in their lives. Those who voted in recent elections have consistently supported measures to weaken entrenched officials, narrow the scope of their responsibility (even of those they voted for), and make government more responsive to their views and demands. Hence the dramatic rise in ticket-splitting, term limit and balanced-budget amendments, public opinion surveys, and traffic in faxes, phone calls, and computer messages to the White House and Congress. To get elected, even incumbents feel obliged to run against the government and in favor of cutting their own authority.

To pressure elected officials into doing the job the voters want, people have taken to making their voices heard in record numbers even in the intervals between elections. So many members of the public picked up their phones and called and faxed the White House to protest Clinton's first nomination for attorney general and his proposal to admit gays into the military during his inaugural week that the venerable White House switchboard collapsed from overload. To be sure, many of those calls were orchestrated by political interest groups and stimulated by irate talk show hosts. But millions more were spontaneous. It was not enough that people had just voted in the election, they wanted their newly elected president to know exactly what they thought of his actions and what they expected him to do in the day-to-day job of running the government.

The public's increasing demand for fundamental change in the political system is far from a temporary phenomenon prompted by an emotional reaction to a few highly charged issues like gays in the military or health care reform. It represents a permanent change, made possible, at least in part, by newly empowering interactive telecommunications technologies and by the speed of information that these technologies have spawned.

In a *New York Times* column, Russell Baker—only somewhat facetiously—attributed the record-breaking Republican triumph in the 1994 election, giving the party control of both the House and Senate, to the voters' new sense of personal empowerment. That empowerment, Baker wrote, had been stimulated by the invention of the TV remote control, "the ultimate weapon" of the public's liberation. From zapping television channels at home, people moved on to the voting booths and zapped incumbent Democrats. "Channel-surfing gave millions a taste of power over an old tyrant [television]." It made them "feel comfortable with the idea [that they] . . . could zap a politician" as well. Baker's explanation was not by any means far off the mark.

THE MODEL OF THE FINANCIAL MARKETS

To get a sense of how telecommunications technologies are transforming the structure of American politics, it may be helpful to look at what they have already done to transform the structure of the world's financial markets. Every day, in trading rooms throughout the globe, traders sitting in front of computer screens pass judgment on different governments' policy decisions and fiscal moves. They tap out electronic "buy" and "sell" orders on their terminals, which are transmitted via instantaneous worldwide telecommunications data networks. Cumulatively, these "yes" or "no" votes serve as an immediate and very public series of financial plebiscites on government policies.

Governments, in turn, have learned to react quickly and decisively to the instant verdicts of their financial constituencies and change major economic decisions and monetary policies overnight when necessary. In the summer of 1993, huge pressure from the world's money traders suddenly forced European central banks to reverse course and stop propping up weak currencies. As one banker put it, "No matter what political leaders do or say, the screens will continue to light up, traders will trade, and currency values will continue to be set . . . by global plebiscite."

The world of finance has already adapted to an environment of instantaneous voting by interactive telecommunications technologies. These devices process and transmit information to and from many sources over great distances at the speed of light. The world of politics is next. Although slower, more complex, less flexible, and more resistant to change than the financial marketplace, the political marketplace will have no choice but to adapt to the rising pressure of public opinion transmitted through increasingly widespread use of telecommunications. The process has already gotten under way, although this country's traditional political institutions—the parties, unions, civic associations, election, and primary systems and even the presumably all-powerful mass media—remain mostly ill-prepared to cope with the radical changes to be brought on by the telecommunications transformation that is overtaking them.

In the financial marketplace, "Intellectual capital is becoming relatively more important than physical capital . . . the new source of wealth is not material, it is information, knowledge. . . ." There is a corollary in politics: For the electorate, information and knowledge are now key, and the new source of political power lies in their widespread dissemination through telecommunications.

THE ROLE OF THE MEDIA

A wonderfully perceptive, if bleak insight into the long-range impact of the telecommunications media on American democracy was offered by a reader of *The New York Times* in a letter published on September 2, 1993. The most "fitting analogy for television," suggested Annamarie Pluhan of Ellicott City, Maryland, "might be the ancient Roman waterworks. A sophisticated technology brought running water into private homes, public bathhouses and imperial palaces. The aqueducts were a great accomplishment." But the Roman aqueducts' pipes were made of lead and slowly, imperceptibly the population was poisoned. "Some historians consider lead poisoning a major cause of the fall of the Roman Empire. Future historians may look back on television in the 20th century as the pollutant that caused the failure of American democracy."

Modern communications technology, by itself, will be neither democracy's savior nor its terminator. But unquestionably it will continue to have enormous influence, both for better and for worse, on the nature and character of our political system. Canadian media guru Marshall McLuhan argued that all history has been shaped by the dominant communications media of each era. Oral communication, McLuhan wrote, was the mark of closely knit, myth loving, superstitious, resistant-to-change tribal societies. Writing/printing introduced reason and linear thinking and made science and democracy possible. Electronic communication, an extension of our senses, is, according to McLuhan, in the process of transforming our society into a "global village" that shrinks geographical boundaries and transforms diverse people through shared experiences and the instantaneous passage of news.

McLuhan's media interpretation of history is echoed by political theorists who argue plausibly that the development of papyrus, the first portable medium of communications, was instrumental in making it possible to create and then govern the far-flung Egyptian and Roman empires. The invention of movable type and the spread of literacy stimulated a commercially based society, broke the monopoly of the medieval church, and fed the Protestant Reformation. The Linotype machine and high-capacity rotary press produced mass circulation newspapers, which—until radio and television came along—largely replaced personal-encounter politics. The gospel for our present-day society became the information and "wisdom" delivered by the press-politics-commerce connection that came to the fore. Today, interactive telecommunications media are bringing about still another major societal transformation, reviving personal-encounter politics and reintroducing a variant on ancient direct democracy, vastly extended through electronic communications.

The charming "global village" that McLuhan envisioned in *Understanding Media: Extension of Man* and other books is the product of the telecommunications revolution. However, it is not in reality a homogeneous, tightly knit electronic world community. Instead, the global village appears to be shaping up as a quarrelsome, heterogeneous, tribal society with clashing ethnic, political, and economic interests and highly diverse ideological factions. In the spring of 1992, I experienced a vivid personal demonstration of that increasingly apparent fact during a session of the first International Congress on Cancer, AIDS and Society at the UNESCO headquarters in Paris. Sponsored by UNESCO, the European Community, the World Health Organization, and the government of France, and organized by the International Council for Global Health Progress, on whose board I served, the congress sought to address the worldwide medical, social, economic, and political issues of these two raging epidemics of our time.

I spoke at the session devoted to communication issues, which, to my surprise, attracted a standing-room-only crowd. Communications were of passionate concern to the international medical experts, health professionals, government officials, journalists, policy specialists, and observers gathered at the Congress. The first to speak in our session was Dr. Arnold S. Relman, professor of social medicine at Harvard

University and for many years the distinguished editor of the *New England Journal of Medicine*. Dr. Relman skewered the popular press for its simplistic, often sensational, and scare-driven reporting of both cancer and AIDS. In the media's relentless competition for public attention, he charged, both the good and bad news about the diseases were consistently overstated and overdramatized. It raised people's hopes by exaggerating the significance of experimental and preliminary research data, with every new medical finding reported as a "wonder," a "miracle cure," and a "dramatic breakthrough." Or the press terrified its readers and viewers by apocalyptic reporting of "plagues," "epidemics," and "mysterious fatal diseases."

Before airing or publishing news and features about new developments in cancer and AIDS, Dr. Relman urged, television, radio, newspapers, and the mass magazines should always clear their stories with authoritative professional journals, like the *New England Journal of Medicine*, or with responsible government agencies, like the Federal Drug Administration and the National Institutes of Health. The press should never report experimental treatments and preliminary research findings, he insisted, unless they first had the stamp of approval of the proper authorities. The doctors in the audience applauded.

As the panel's television representative, I followed Dr. Relman to the podium. I had no inclination to take issue with his denunciation of the media's reckless and irresponsible reporting of the two diseases. Dr. Relman's sharp criticisms, I had to agree, were all too justified. But I *did* take issue with Dr. Relman's insistence that the press should withhold from the public all cancer and AIDS information that was still experimental in nature and that had not been cleared by proper authorities for general use. The good doctor was being far too cautious, I argued, in wanting to limit the public's access to medical information to only what has been professionally sanctioned and officially approved.

In fact, I had a vested personal interest in the issue, as I confessed to the international audience in the UNESCO hall. I had just completed a year of invasive surgical procedures and intensive radiation therapy for a rare form of cancer, which, in fact, had been misdiagnosed by one of the leading medical institutions in the United States. I was not so trusting of official authority or respectful of professional experts. The one lesson I took away from my own experience with cancer was the importance of taking the final decisions about my treatment into my own hands. As a patient, I wanted to know what my choices were. I did not want to be kept in the dark if promising new therapies existed that were still experimental. No officials, experts, or medical professionals should be able to stop patients and their families, the people most directly affected, from learning all they could about their disease, including the alternatives for treatment.

With the proliferation in the United States of on-line computer information services and increasing access by patients to medical data formerly available only to medical professionals, the restrictions Dr. Relman proposed would not work in any event. The new telecommunications and computer technology is changing the traditional relationship between members of the public—at least those with access to personal computers—and authority figures of all kinds. It is arming the former with information once unavailable to the public at large, and putting them in a better position to ask questions, challenge experts, and participate in the decision making—not unlike the enormous changes wrought by the invention of the printing press centuries earlier.

Of course, I argued, the press should keep experimental and preliminary successes in proper perspective. But when covering medicine, as when covering politics, the worst thing the press could do would be to consistently defer to official orthodoxy and professional authorities. Doctors, like government officials, are not gods. The more patients and their families know, the more informed and intelligent their own judgments can be. The journalists in the audience applauded my rebuttal.

I was followed on the panel by a fiery young newspaper reporter from Central Africa, who promptly assailed both Dr. Relman and me. Central Africa, he said, was being ravaged by AIDS. The epidemic had already taken a terrible toll in his country. Yet his government stubbornly refused to admit that AIDS was even a problem. At home, the subject was taboo, covered by a blanket of silence. All mention in the popular press was forbidden, even though the people of his country knew that something had gone terribly wrong because they could see for themselves that so many were being felled by a mysterious and devastating illness.

The African journalist made an eloquent plea to the world community to push for unfettered, uninhibited, even lurid coverage of AIDS in his country and every country. As far as he was concerned, the more sensational and shocking the stories, the more useful they would be. He favored headlines that shouted out AIDS PLAGUE, AIDS DISASTER, GOD'S PUNISHMENT, MYSTERIOUS DEATHS, NO CURE, to frighten his fellow countrymen and -women into changing their destructive social behavior that only accelerated the spread of AIDS. Looking straight at Dr. Relman and me, who remained on the speakers' platform, the newsman said he cared not a whit about our complaints of irresponsible reporting, exaggerated stories, the premature release of experimental cures, or the need for validation. For his country, all of that was "irrelevant dithering." He preferred coverage considered irresponsible and exploitative by Western standards. His compatriots from African and Latin American nations applauded him enthusiastically.

Reflecting on that global experience, it was clear to me that all three of us were right—in our own contexts. Different people in the global village will see, hear, and interpret exactly the same information in entirely different ways, depending on their individual needs, outlooks, and backgrounds. As we move into an era characterized by more, though not necessarily better, media alternatives and instantaneous transmission of accurate as well as inaccurate information, the likelihood is that the future inhabitants of the global village could grow more fragmented, divisive, and factionalized rather than less so.

McLuhan viewed the new media as the epicenter of society, expressed in his widely circulated aphorisms "The medium is the message" and "The medium is the mass age." A less catchy but far more realistic and practical media perspective was expressed to me in the course of a conversation with my own doctor, Shirley Grossman, a favorite aunt and extraordinarily wise and sensible physician who attends patients in the Bronx's Montefiore Medical Center. Her comments about the transformation of the telephone, one of contemporary society's most potent older communications media, were triggered by a momentary interruption from the click of a call-waiting signal: "You know, amazing things have happened to the telephone in the past few years," Dr. Grossman said when our conversation resumed. "I still remember the simple, sturdy dial telephones we used to use. They were all black. I also remember talking to live telephone operators. They were all women and seemed to know everything. Now we have call-waiting options; fancy new multicolor Touch-Tone phones of all shapes and sizes with memory, recall, and automatic dialing; video phones, and phones attached to computers, and so many other time-saving devices, most of which I either can't figure out how to use or really don't need. Suddenly, there are so many different telephone companies, trying to sell me so many different packages, I don't have a clue which one to choose.

"Also, people use the phone today in the strangest ways. I get calls from patients who are in the bathroom, or walking on the street, or driving a car, or even flying in an airplane. I get fax messages printed out through my phone line in the hospital. I get messages from one person while I'm still talking to another. I get dialed up by a whole bunch of different doctors in different places, who can talk to each other all at the same time. I can reach people in Tokyo, Paris, or Tel Aviv as easily as if they were in the apartment next door. I see ads offering answers to crossword puzzles, medical diagnoses, sex talk, stock quotes, baseball scores, and God knows what else over the phone. I am told to buy anything I want by telephone.

"Every night at dinner somebody I don't know calls me to sell me something I don't want. When I listen to talk shows on radio or television, I hear people from all over the world calling up Larry King, famous radio and television stars, or even the president of the United States while everybody else listens in. Telephone surveys ask me to give my opinion about health care reform, taxes, or the president's State of the Union message. But when you come right down to it, have all these telephone bells and whistles made us any better or smarter? Have they raised the level of conversation or improved the quality of our knowledge about important things? Absolutely not! None of this new telephone technology has made people any more articulate, informed, or intelligent than they used to be. Despite all these automatic devices, people have no more time than they used to. Once we would spend money to call someone only when we had a good reason to; when we had something important or useful to say. Now everyone uses the telephone all the time for no particular reason, just because it's there."

The medium does affect the message. And the new media are profoundly transforming our political system. But Dr. Grossman is right. Whether it will be better or worse depends entirely on how the new telecommunications technologies, and the information they transmit, will be used, by whom, and to what end.

THE IMPERATIVE FOR
SOUND PUBLIC JUDGMENT

The nation's citizens today are being overwhelmed by extraordinarily difficult, seemingly intractable political, social, and economic problems—the high cost of health care, epidemics of crime and drug abuse, faltering public education, lack of job security, pervasive pollution, increasing inner-city poverty, racism, and the breakdown of family life. Well-informed, experienced political elites have made little or no progress solving any of these problems. Out of increasing frustration, some members of the public at large are withdrawing from any participation in politics. But some are beginning to take advantage of their new ability to make their voices heard. Through their rising use of the new mechanisms of direct and interactive communications, many citizens who have not withdrawn in frustration and disillusionment are beginning to have an active and important influence on the making of day-to-day political decisions.

In the electronic republic, as citizens at large gain the power of self-governance, the need to inform and educate the American people about these complex issues and about the workings of our political system has become, if anything, more important than ever. In the words of one political theorist, we need to stem "the decline of a genuine politics in the modern age and to suggest what a form of political association built upon plurality, sheer human togetherness, might look like."

It is obvious that an informed and interested public is the key to successful self-governance. With the public at large playing a critical part in the government's decision-making process, it is essential that the public know what political alternatives are available and what their costs and consequences will be. New methods and new systems must be devised to enable ordinary citizens to reach responsible and informed judgments. Since information has become "society's main transforming resource," the public's ability to receive, absorb, and understand information no longer can be left to happenstance.

No military general would willingly send his army into battle untrained and ill-prepared, no matter how well-equipped. Yet today, the American public is going into political battle armed with increasingly sophisticated tools of electronic decision making but without the information, political organization, education, or preparation to use these tools wisely. Information about public issues can be made immediately accessible to every citizen in a form and at a level to suit each individual's needs. Without a conscious and deliberate effort to inform public judgment, to put the new interactive telecommunications technologies to work on behalf of democracy, they are more

2

THE ROOTS OF THE ELECTRONIC REPUBLIC: DEMOCRACY'S THIRD TRANSFORMATION

Today's telecommunications technology may make it possible for our political system to return to the roots of Western democracy as it was first practiced in the city-states of ancient Greece. Tomorrow's telecommunications technology almost certainly will.

What distinguishes democracy from every other form of government is the principle that the rulers are subject to control by the ruled. In a proper democracy, power flows from the bottom up rather than from the top down, an unnatural course that defies the law of gravity, which may explain why democratic governments have been so rare in history and their rule so precarious. Those in power, when they have police and armies to back them up, do not relinquish their authority readily, especially to those they rule.

The fall of tyrannical communist regimes throughout Eastern Europe and the universal demand for their replacement by some form of democracy originally led many to assume that an era of widespread democratic governance was at hand. Few recognized how difficult a system democracy is to create and maintain. Today, as we observe the wrenching struggles that all new democracies experience throughout the world, we begin to understand how much needs to be done to nurture and cultivate democratic institutions and principles. "Although in our time democracy is taken for granted, it is in fact one of the rarest, most delicate and fragile flowers in the jungle of human experience. It existed for only two centuries in Athens and less than that in a small number of Greek states."

The United States today, obviously, has almost nothing in common with the tiny city-states of ancient Greece. Citizenship in Greek democracy was highly exclusive rather than inclusive, as modern democracy has become. The citizens of Athens depended on the labor of slaves, the work of women, and the exploitation of servants and aliens, none of whom had any political rights, to gain the leisure to govern themselves. But if, in truth, the United States is moving toward a new modern-day form of direct democracy, made possible by its commitment to universal political equality and propelled by new telecommunications technology, then it would be useful to examine the roots of direct democracy. And they lie in ancient Greece.

DIRECT DEMOCRACY: THE FIRST TRANSFORMATION

In the very first form of democracy, the rulers and the ruled were the same. Every citizen of ancient Athens could take part in making every decision of state. When democracy was born some twenty-five hundred years ago, the idea of *representative* government as we know it today did not exist. There was no need for it. The approximately forty thousand citizens of Athens who wanted to involve themselves in civic affairs simply attended the assemblies where the issues were discussed and the decisions made. The citizens present decided by a show of hands what should be done. They also took turns administering the city-state, drawing lots to settle who would take on which job. While Athenian citizenship was highly exclusionary—only adult males of native parentage were qualified to be citizens, which excluded 90 percent of the inhabitants—all those who *were* citizens were entitled to participate in governing the city.

The small Greek city-states were bordered by kingdoms, tribes, satrapies, tyrannies, military states, and miniature empires. In the classical Athenian plays, dialogues, and histories that survive, the citizens of Athens were obsessed with the idea of power and politics. They talked about the subject endlessly: What is the nature of power? Who has it? Who uses it and misuses it? How should it be administered? How should it be controlled? Where does it spring from? Are the gods who have power just? Are moral imperatives stronger than political power?

What emerged from these discussions was the idea of the polis, the civic community that was the center of the life of Athenian citizens outside the home. The polis organized and fulfilled the Athenians' social relations and provided "the only significant link between citizens beyond the home. . . . To the Greeks . . . 'political' meant the same as 'common' . . . and referred to what concerned everybody." Politics became the dominant element in the life of their community. "[T]he citizen's affiliation to the polis became more important than any other."

Athenian democracy thrived because the exercise of citizenship was viewed as central to the life of society. The one endeavor in which all Athenian citizens were considered equal was their capacity as citizens. In ancient Greece "everyone, not only the prominent members of society, but also, to an extraordinary extent, whole communities, and in Athens even the common man—acquired a certain stature in relation to whatever happened in the world, an ability to influence the course of events that is scarcely conceivable today. They had a fairly direct, concrete, and existential share in the making and execution of decisions (and could identify themselves with the decision-makers . . .). Whatever happened . . . they were close to events and inextricably involved in them." Politics became the major preoccupation shared by all, rather than the concern of the ruling few.

The Athenians understood that no lasting improvement of the civic order would be possible unless broad sections of the citizenry assumed active political roles. "To this end, certain sections of the political intelligentsia allied themselves with the broad mass of the citizens to promote a trend toward wider participation within the polis." In the direct democracy of ancient Athens, all citizens, no matter how rich or poor and regardless of their social status, "were deemed capable of judging what was in the interest of the community, and therefore all were responsible."

Through the hard work and concerted efforts of the leading citizens and the rank and file, the Greeks organized their political institutions to foster widespread citizen participation. The assemblies met frequently, an estimated forty times a year and on other occasions when necessary, to discuss and decide the issues of the community. The citizens of ancient Athens would trek to the Pnyx, the hill near the Acropolis that served as their meeting place, to exercise their responsibility to govern themselves.

Athenian citizens based their concept of direct democracy on the logical but rarely practiced principle that nobody is as capable of making judgments in a person's interest as that person himself. The members of a community, rather than some self-appointed higher authority, are in the best position to understand and act upon their own common interest. Not all Athenians subscribed to that principle and agreed with the idea of democratic self-rule. Both Aristotle and Plato, our major sources of information about the golden age of Athenian democracy, were deeply critical. Aristotle feared that democracy would inevitably drift downward to the unskilled, uneducated, boorish mass of ordinary people whom he despised. "The faculty of speech and the life in a polis" should be restricted, Aristotle urged, since they are what "distinguish a Greek from a barbarian, the free man from the slave." Unlike barbarians, Greeks "conducted their affairs by means of speech, through persuasion, and not by means of violence, through mute coercion."

Plato disapproved of the very idea of democracy, arguing that proper government required the trained skills of a uniquely competent corps of professional guardians, an elite civic priesthood whose members should be selected at an early age and rigorously prepared for their role. Both Plato and Aristotle, whose ideas America's Founding Fathers studied closely, believed in a ruling elite, viewing rulers as authority figures like parents or teachers, and the ruled as subservient children and pupils. Like children, members of the public at large, they argued, are too ignorant, unsophisticated, and immature to govern themselves.

To function effectively, the direct democracy of ancient Greece required a political unit small enough in size so that all citizens who wished to attend the governing assemblies could do so, small enough in population so that everyone could deliberate on the issues together and then resolve them by a show of hands, and sufficiently homogenous and self-contained so that irreconcilable interests and competing factions would not prevent the assembled group from reaching consensus.

The citizens of Athens recognized that responsible citizenship would not come about automatically; it had to be carefully cultivated. A Council of 500 was formed—consisting of about one citizen in sixty —to hold preliminary deliberations on major matters due for public debate and to prepare the way for all citizens to make sound judgments in the general interest. It was the responsibility of the council to supply information to all the citizens and put forward proposals. The council was designed to make sure "that it had as few rights of its own as possible and that its membership did *not* include the most influential and experienced men (who might have . . . monopolized politics)." The council's members did not act on behalf of the body of citizens; their job was to help prepare the citizens to act for themselves.

The council's members were chosen by lot; no one could serve for more than two terms, and expenses were paid so that even poor citizens could take part. The job of preparing citizens to deal with whatever civic issues would come before them was considered so important that a tenth of the council members, the so-called *prytaneis,* were required to be in constant attendance in the marketplace. Some even were required to sleep there.

"A further requirement was that . . . broad sections of the citizenry be able to form judgments and evolve a concerted political will in important matters. It was therefore necessary to create opportunities for extensive participation in politics. Politics had to be brought into the lives of the citizens. It was not enough for them to have the necessary information and an overall view of polis affairs: they must be able to exercise an effective influence, commensurate with the effort they invested in it. In particular, communication had to be established between different parts of the community throughout the country. There had to be meeting points for the exchange of ideas, institutions that would create links among the citizens. It was not enough 'to bring the citizens to the polis': they had to 'come together' in new ways. . . . Not only must politics ultimately become the affair of the citizens, but the affairs of the citizens must become the stuff of politics. The content of politics must change."

Protagoras described "the education of Greek citizens to political virtue as a broad process that is promoted variously by parents, fellow citizens, and polis institutions." Indeed, "just as all citizens are brought up to speak the language, so too they are educated to political virtue."

"Only among us," said the great Athenian leader Pericles in his famous Funeral Oration, "is a man who takes no part [in political affairs] called, not a quiet citizen, but a *bad* citizen." As we said, to sustain their intense, all-absorbing civic life, the ancient Greeks relied on the labor of aliens, slaves, and women to do the day-to-day work of their society. They were excluded from citizenship and political life. As Hannah Arendt described their concept of citizenship, "In Athens and throughout antiquity up to the modern age, those who labored were not citizens and those who were citizens were first of all those who did not labor or possessed more than their labor power." Still, by introducing the world's first government of citizens, by citizens, and for citizens, the ancient Greeks brought about an unprecedented transformation from what until then had been rule by the very few to rule by the many.

It says much about the limits of the human capacity for self-government that political systems based on political equality and self-rule have been so rare and have affected so few. For Athenian citizens at least, theirs was a society governed by persuasion and consensus as opposed to force and coercion. It was a radical departure from the past, an interlude of democracy in an otherwise unbroken line of authoritarian rule.

We have a good deal to learn from the way the ancient Athenians fostered their democracy and cultivated civic participation in their community. But today it might be argued that the urgent pace and complexity of modern American life make such dedication to the public good all but impossible, except for the very rich and for full-time professional politicians. In addition, competing distractions and entertainments are so readily available on television and elsewhere in our abundant consumer society that it would be highly unlikely that ordinary citizens would immerse themselves in civic life, even if they could find the time to do so.

The Greeks recognized that civic responsibility, political participation, and the widespread dissemination of information about public affairs could not be left to chance. They worked hard at every facet of developing an informed citizenry. Like them, we need to invent new ways and create institutions to cultivate good citizenship, to encourage informed public participation, responsible public service, and sound public judgment. In principle, the problem of how that can be accomplished is not all that different from the challenge that faced the citizens of ancient Athens except that, unlike us, the Athenians, the first to practice democratic governance, had no precedents to guide them.

Direct democracy lasted in Athens for approximately two centuries, with a few brief intrusions by demagogues or overly ambitious generals who sought power and then were quickly deposed. Democracy died after a period of incessant wars, imperial expansion abroad, and the rise of demagoguery at home. During the time that Athenian democracy flourished, a delicate balance had been struck between the power of ordinary citizens and the power of the rich. Whatever the social or financial disparities between the classes of citizens, they were considered equal in the ·public sphere, which in ancient Athens was considered the most important realm of life. Later, however, as the rich grew richer, power gravitated away from the popular assemblies toward the wealthy few. In return for taking on the job of making the major decisions of state, the richest citizens were expected to volunteer to contribute sizable amounts of their own money for the public benefit. In the end, those decisions were made in the wealthy citizen-contributors' own interest rather than for the common good, and Athens fell into decline and decay.

Direct democracy resurfaced for a period in the ancient republic of Rome and later still in the vibrant Renaissance city-states of Florence and Venice before disappearing altogether for many centuries. It was a governmental system that could work well enough in the small, self-contained, homogeneous communities of the Mediterranean basin. But direct democracy could not survive the size and complexity of the pluralistic, sprawling nation-states that developed in modern times. The world reverted to rule by kings and emperors and hierarchical, authoritarian, and despotic regimes, which have characterized every government on every continent for most of recorded history.

REPRESENTATIVE DEMOCRACY: THE SECOND TRANSFORMATION

Not until the end of the eighteenth century, with the American and French revolutions, was the flame of democracy finally rekindled, although in a form that would have been totally unrecognizable to the citizens of ancient Athens. Rather than rule themselves as the Athenians did, the Americans entrusted the decision-making power of government to a select group of elected officials, chosen to rule on their behalf. What emerged in the infant United States was the world's first representative republic, democracy's second transformation. What distinguished representative government in the United States from the earlier cantons of Switzerland and parliaments of England was the fact that the electorate, not the elect, *the people*—not the government, the monarch, or even the parliament—have sovereignty. That is a comparatively recent idea in the long span of history.

In 1644, English poet–political philosopher John Milton's seminal *Areopagitica* laid the foundation for the then brand new idea of political representation as a way to include all the people. "The ruler is the people," Milton wrote, "an actively self-governing people . . . conducted not among an elite of grace or virtue, or within the walls of a representative parliament or ecclesiastical assembly but among 'the whole nation.' "

By proposing the idea of representation, Milton advanced the concept of democracy from small, ancient, self-contained communities of a few thousand citizens to the then-emerging nation-states, embracing millions of citizens. By choosing those who would govern them, the people would also, in effect, be governing themselves. To Milton, "the collective rationality of individualism" would provide the order necessary for political survival even in diverse and pluralistic societies. "While the people are not infallible, at least they are more likely to get at the truth of a matter than any other claimants to authority."

Milton's *Areopagitica* anticipated England's "Glorious Revolution," in which Parliament's army, under the leadership of Oliver Cromwell, overthrew and executed Charles I in 1649 and by so doing, also killed the idea of the divine right of kings. After Milton, John Locke, publishing his *Two Treatises on Government* in 1690, proposed that government should rule by the consent of the majority, which might be given "either by themselves or their representatives chosen by them." But like Milton, Locke had little to say about how representation actually might work and about its place in a democratic structure of government. In the eighteenth century, the radical and passionate French political philosopher Jean-Jacques Rousseau sought to undermine the emerging idea of representative democracy on the grounds that it was not populist enough. "The instant a people allows itself to be represented it loses its freedom," he wrote in his *Social Contract*. "The English people thinks it is free. . . . It is free only during the election of members of Parliament. As soon as they are elected, it is a slave, it is nothing."

Rousseau's jaundiced view of the British Parliament was very much on target. In Britain, the monarchy had ceded power not to the people, but to the parliament, which "was the supreme lawmaking power." According to Blackstone at the time (1765), "that absolute despotic power, which must in all governments reside somewhere," was entrusted to Parliament, whose members therefore had "sovereign and uncontrollable authority" to make or unmake all laws. That remains true in England to this day. In contrast to the United States, where— as James Madison put it—"the people, not the Government, possess the absolute sovereignty," in England, "Parliament is sovereign, and the people have no rights in law if Parliament restricts their freedom."

In 1748, the Baron de Montesquieu recognized that in a widely dispersed, large, and heterogeneous nation-state, which France then was rapidly becoming, it would be impossible for all of the people to assemble in a democratic governing body in the fashion of the ancient

Greek and Italian city-states. Montesquieu urged instead that the people should choose a manageable number of representatives to do for them what they could not do for themselves. At the time, Montesquieu's proposal for government by elected representatives was a radical thought. But it struck a responsive chord in France, Britain, and especially in Britain's American colonies across the Atlantic Ocean.

The Americans gave birth to the seminal idea that the people themselves were the source of supreme political power and ultimate political authority. Instead of government being derived from a higher authority—the divine rule of monarchs, the power of tyrants, the coercion of generals, or the force of superior intelligence and ability of a guardian elite—sovereignty, the authority of government itself, came from the ruled, the people underneath. Government was not imposed from above but authorized from below. The people have imprescriptible rights and the government must be their servant, not their master. That fundamental idea stood the traditional theory of governance on its head by embracing a revolutionary political philosophy—representation from the bottom up.

Under the American Constitution, every level of government derives its powers from the people. Government rules only by leave of the governed. According to that theory, the sovereign people initially authorized the formation of the government and defined their individual rights as matters of fundamental law. Any rules that breach that fundamental law, even if passed by Congress, state legislatures, or any other branch of government, are considered to be invalid both morally and legally. "Above these inferior law-making bodies was a sovereign power which had authorized them and which watched over them and could intervene to correct them. The sovereign lawmaking power is the people." Over the years, the process of democratization has been extended to include all the people, not just a select few, to the point that today every adult citizen has the right to vote and participate in government, regardless of race, sex, age, religion, or financial condition.

This underlying belief in the people's sovereignty has both foreshadowed and propelled what is now the third transformation of democracy, the emergence of the electronic republic. Since the people are sovereign, it follows that the more power they have to control their government and to involve themselves in making its decisions, the better.

In the turbulent years that led to America's Declaration of Independence in 1776, the colonists' anger at having no say in their own governance—"No Taxation without Representation!"—precipitated their revolt against British rule. The British, in an effort to co-opt their rebellious subjects abroad, had tried to convince the distant American colonists that they did indeed have representation in Parliament through the doctrine of "virtual representation." According to that doctrine, once seated, every member of Parliament was duty bound to represent not only his own home district but also all subjects of the British crown, including those, like the American colonists, who had no vote. Not surprisingly, the doctrine of "virtual representation" seemed entirely implausible to the Americans and the War of Independence ensued.

Early in the nineteenth century, America's new and unprecedented representative republic, which was rooted in the sovereignty of its citizens, was hailed as, "the grand discovery of modern times" and "the invention that makes democracy . . . practicable for a long time and over a great extent of territory." Ironically, it was also seen as the most effective way to protect the new nation against the excesses of a popular majority while still preserving the underlying ideals and the appearance of citizen sovereignty. Because the new republic embraced such a vast territory and widely diverse population, no single majority faction would be in a position to seize power. And just to be sure, numerous federal checks and balances and a constitutional separation of powers would serve to prevent majoritarian government and mob rule.

While the U.S. Constitution embraced the then revolutionary idea of the sovereignty of the people, the Founders at the same time also found it prudent (and not inconsistent) to severely constrain the role of the people in carrying out the business of government. The only federal officials to be elected directly by the citizens themselves were the members of the House of Representatives. And even that limited voting privilege was at first confined to less than ten percent of the population. Like the ancient Greeks, no women, slaves, indentured servants, aliens, or non-property-holding poor could vote. None of the highest officers of the land was popularly chosen. Instead, an elite electoral college was created to select presidents and vice presidents; U.S. senators were named by state legislatures, and federal judges were appointed for life. Moreover, before any law could be put into effect, it had to be filtered through a complex series of checks and balances —passed by two separate houses of Congress, each chosen in a different manner; approved by the president; and finally, subject to Constitutional review by the courts. The latter's power of judicial review gave the federal officials farthest removed from the people, justices appointed for life, the final say.

THE FEDERALIST
REPUBLICAN STRAND

American government since its inception has been designed with layers, filters, and barriers to restrain the power and passions of the popular majority. Under the U.S. Constitution, the actual participation and involvement of the general public in the federal government was extremely limited. Even for the small minority who were eligible to vote, their active exercise of citizenship, in effect, only began and ended on election day. As George Will, among others, has written, "The point of representative government is that the people decide . . . who shall decide." Unlike in ancient Athens, the citizen's role was not to govern himself, but at best to elect those who would be most competent to govern on their behalf, those who possess what British liberal philosopher John Stuart Mill called "the acquired knowledge and practiced intelligence of a specially trained and experienced Few." The new Constitution had put in place a modern-day method of selecting Plato's guardians.

The drafters of the Constitution solved the practical problem of merging thirteen independent states into one nation by creating a representative federal republic instead of a single central government. It was an ingenious political solution that embraced the ideal of popular sovereignty and at the same time also served to insulate the new nation from "an unqualified complaisance to every sudden breeze of passion, or to every transient impulse" of the people the new government was designed to serve. The Federalists, led by Hamilton, were primarily concerned with protecting life, property, and public order by reining in the unbridled power of the majority. Among the Constitution's drafters in Philadelphia, that was their principal driving force: "Almost unanimous was the opinion that democracy was a dangerous thing to be restrained, not encouraged, by the Constitution, to be given as little voice as possible in the new system, to be hampered by checks and balances. [Massachusetts's Elbridge] Gerry declared that the evils the country had experienced flowed from 'the excess of democracy.' [Virginia's John] Randolph traced the troubles of the past few years to the 'turbulence and follies of democracy.' Arguing in favor of a life term for Senators, [New York's] Hamilton exclaimed that 'all communities divide themselves into the few and the many . . . , the mass of the people who seldom judge or determine right.' [New York's Gouverneur] Morris wanted a Senate composed of an aristocracy of wealth to 'keep down the turbulence of democracy.' Madison, discoursing on the perils of majority rule, stated that their object was 'to secure the public good and private rights against the danger of such a faction and at the same time preserve the spirit and form of popular government.' "

In sum, the Constitution organized things so that the people's representatives would not be able to transform public opinion raw into public policy. As Hamilton explained in the Federalist Papers, "It is the duty of the persons whom they have appointed to be the guardians of those interests to withstand the [people's] temporary delusion, in order to give them time and opportunity for more cool and sedate reflection." Nevertheless, however carefully constructed the new government was to withhold power, insulate the country against the people's "temporary delusion," and provide the "opportunity for more cool and sedate reflection," American representative democracy was firmly rooted in the principle that human beings in general "possessed the inherent capacity to govern themselves." As James Madison put it, although men differed greatly in their "faculties," they all had "reason" sufficient to entitle them to live a free, republican life. They would do this *not* by making the major decisions of government themselves, as in the direct democracies of ancient Greece, or by means of modern-day referenda, plebiscites, and ballot initiatives, "but by a system of election and office-holding which subjected the representatives to close and continuous supervision by the represented." In that manner, the American people were given at least an indirect role in deciding what was in their common interest.

THE ANTI-FEDERALIST
POPULIST STRAND

Although the efforts of the men who wrote the Constitution were, in Tocqueville's words, "directed against this absolute power" of majority rule, the new nation also included a strong and radical populist strand whose roots lay in the American Revolution. This movement's best-known spokesmen, Tom Paine and Patrick Henry, championed a far more inclusive view of democratic government than the Constitutional Founders offered. Paine's earlier *Common Sense*, the stirring call to revolution published in January 1776, originally set the theme for the populist vision: "In this first parliament, every man by natural right will have a seat," Paine wrote. "A representative assembly should be in miniature an exact portrait of the people at large. It should think, feel, reason and act like them." He viewed government solely as an instrument of the people that should never be given the power to act beyond the people's will. *Common Sense* was the biggest seller of the day in the American colonies.

Thomas Jefferson, also an anti-Federalist but less radical than the others, refrained from going public with his initial opposition to the new Constitution. Jefferson was in France while the document was being drafted. His ideal was to transplant the classical Athenian model of direct democracy to the new world. Jefferson envisioned an America of independent yeomen-farmers in small, self-contained political communities who would deliberate among themselves, decide their own affairs, and govern themselves, much as his beloved classical Greeks did in ancient times.

The fathers of American populism, contrary to the Constitution, wanted to bring a form of participatory democracy to the New World, a political system that would focus on the interests and prosperity of the common people. They feared, rightly as it turned out, that the new central government would exclude the vast majority of citizens from achieving the kind of direct political control over their own affairs that had been their main reason for starting the War of Independence in the first place.

Concern that the proposed new central government would not speak for the people was so intense that a provision was included in the original draft of the Bill of Rights that would have required members of Congress to follow the instructions of the majority of their constituents in voting for any new measure. Four of the original thirteen states—Massachusetts, North Carolina, Pennsylvania, and Vermont—already had such mandates in their own constitutions. The provision was ultimately defeated in the House of Representatives only after a compromise was struck that added to the wording of the First Amendment the right of the people to petition their government "for redress of grievances."

"The Revolution brought respectability and even dominance to ordinary people long held in contempt . . . in a manner unprecedented in history and to a degree not equaled elsewhere in the world," wrote historian Gordon S. Wood. But after the Revolution, the authors of the Constitution—fearful of the radicalism and irrationality of the "or-

dinary people"—contrived a form of government that was designed
to protect the new nation from actually being governed by them.

Ironically, the anti-Federalists' populist belief in the "absolute
power of majority rule"—their conviction that it should be the common people who participate in running the daily affairs of government—in the end has become the legitimate American mythology, "after two centuries, the most durable expression of . . . [America's]
yearnings and dreams."

Today, echoes of those once radical anti-Federalist views are being
heard not only from populist billionaire Ross Perot and consumer advocate Ralph Nader but also from ardent conservatives, like Kevin
Phillips, Pat Buchanan, and Jack Kemp. Their major current themes
—"Return government to the people," "Limit government to do only
what the people want it to do," and "Replace the unresponsive political elite"—are reminiscent of the pleadings of Patrick Henry and Tom
Paine. Two centuries later, the pendulum has swung toward the Jeffersonian vision of a more inclusive and interactive form of democratic
governance.

THE ELECTRONIC REPUBLIC:
THE THIRD TRANSFORMATION OF DEMOCRACY

The transformation to a modern-day extension of Jeffersonian participatory democracy is accelerating rapidly. The increasing speed of information, the growing commitment to political inclusiveness and
equality, and the public's rising frustration at the poor performance
of the so-called professional politicians are all playing a role. But as
we have seen, it is a trend that, in some respects, has been evolving in
fits and starts since the nation was born. It began with the very first
act of the first Congress, to create a citizens' bill of rights, which went
into effect in 1791. Efforts to broaden public participation continued
with the end of property ownership as a requirement for voting in
1860; adoption of the secret ballot in the 1870s; adoption of the Fif-
teenth Amendment extending the vote to African-Americans in 1870;
direct election of U.S. senators in 1913; granting women the right to
vote in 1920; allowing residents of the District of Columbia to vote
for president in 1961; banning poll taxes in federal elections in 1964;
lowering the voting age to eighteen in 1971; adopting civil rights laws
abolishing literacy and state poll tax requirements through the years,

and in 1994, starting to ease the process for voter registration through "motor-voter" registration and mail ballot laws.

The transformation to participatory democracy has also been helped by the remarkable increase in the speed of information, pushed along by the invention of the rotary press and mechanical typesetting, then the spread of the telegraph, underseas cable, telephone, radio, television, cable television, satellite transmission, computerization, and now, digital convergence. As long as weeks, months, and at times a year or more were needed for news to travel from one end of the country to the other, the public could not readily get involved in making day-to-day decisions of government. Citizens had no choice but to delegate that job to their elected representatives and rely on them to use their best judgment in conducting the public's business.

Today, however, information travels with the speed of light, distance is no longer a significant barrier, and citizens' voices not only are transmitted but also amplified by the power of electronics and the vastness of cyberspace. For example, in 1994, a thirty-two-year-old Spokane engineer named Richard Hartman, who had never been active in politics before, was so angered by what he viewed as the pork-barrel spending provisions of the newly signed crime bill that he tapped out his opinions in a computer bulletin board on the Internet. Instantly, Hartman "received dozens of replies, some asking what *he* was doing to defeat [House] Speaker [Tom] Foley. So Mr. Hartman started a PAC called 'Reform Congress: The De-Foley-Ate Project.' He publicized it on the Internet and talk shows, and raised $26,000 in small checks. The money paid for ads and an anti-Foley car parade . . . that had over 150 vehicles. Mr. Hartman's steering committee consist[ed] of 12 people from around the country, 10 of whom he . . . never met or even *spoke* with. They coordinated their activities on the Internet, which has an estimated eight million regular users and is growing at a rate of 15 percent a month." Along with a good many other Democrats, Speaker Foley lost the election in a close race against a less-well-known opponent. Mr. Hartman's nationwide cyberpolitical blitz undoubtedly contributed to his defeat.

On the opposite end of the political spectrum is "PeaceNet, a liberal computer network, [which] was started in 1985 and now connects 3,000 progressive groups world-wide. 'I can have an issue alert faxed to 900 people overnight,' says Ann Lewis, a Democratic activist who works with Planned Parenthood. 'You no longer have to drag people to meetings.' She says the 'democratization' of political knowledge 'makes it harder and harder to be an insider.' " The conclusion: "Voters . . . have come to believe they have just as much right to propose solutions as do Beltway policy wonks and experts." The information superhighway has put America on the road to the electronic republic.

DEMOCRACY'S FUTURE

Surveying the twenty-five-hundred-year history of democracy, Yale political scientist Robert A. Dahl speculated in his remarkable book, *Democracy and Its Critics*, published in 1981, on the form that the next transformation of democracy might take. In the coming millennium, Dahl predicted, new telecommunications technology will exert a powerful influence for change on the democratic process. The capacity of telecommunications to make information about public issues "almost immediately accessible" in a form and at a level appropriate to virtually every citizen "can be used to damage democratic values and the democratic process, or it can be used to promote them," Professor Dahl concluded. "Telecommunications can give every citizen the opportunity to place questions of their own on the public agenda and participate in discussions with experts, policy-makers and fellow citizens." But, he warned, it will take a "conscious and deliberate effort" to make the new technology work for democracy rather than against it.

THE ROOTS OF THE FUTURE

The citizens of ancient Athens who assembled on the hill of Pnyx would never recognize today's impersonal, far-flung system of democracy in the vast United States involving more than two hundred million people. It has almost nothing in common with their own intimate face-to-face style of direct political involvement involving only thousands. But the next century is likely to tell a very different story. Interactive telecommunications technology makes it possible to revive, in a sophisticated modern form, some of the essential characteristics of the ancient world's first democratic polities. Instead of a show of hands, we have electronic polls. Instead of a single meeting place, we have far-flung, interactive telecommunications networks that extend for thousands of miles. In place of personal discussion and deliberation, we have call-ins, talk shows, faxes, and on-line computer bulletin boards.

A political system that, in the words of John Adams, "depute[d] power from the many to a few of the most wise and good" two hundred years ago has begun to reverse course and depute power to the many in an effort to wrest power from the few (who no longer are seen by many voters as either wise or good). "Every cook has to learn how to govern the state," Lenin declared in 1917, describing how the Bolsheviks planned to retain power. The Bolsheviks, fortunately, are long gone. But today's cooks can help govern the state by communicating their views electronically on the issues before it.

Representative democracy has always been seen as the only form of democratic governance that can work in the large, pluralistic nation-states of the industrial age. Now the United States, as well as other advanced nations, are moving quickly out of the industrial age and into the information age. Interactive information technology has the potential to become the twenty-first century's electronic version of the meeting place on the hill near the Acropolis, where twenty-five hundred years ago Athenian citizens assembled to govern themselves.

The electronic republic cannot be as intimate or as deliberative as the face-to-face discussions and showing of hands in the ancient Athenians' open-air assemblies. But it is likely to extend government decision making from the few in the center of power to the many on the outside who may wish to participate.

3

THE RISING FORCE
OF PUBLIC OPINION

Public opinion is an elusive and highly changeable current that can veer abruptly. In the electronic republic, where decisions on major issues increasingly require direct public input and participation, the need to understand public opinion—to know what people think—has become the central ingredient of politics. As the very first issue of *Public Opinion Quarterly* announced back in 1937, "A new situation has arisen throughout the world, created by the spread of literacy among the people and the miraculous improvement of the means of communication . . . for the first time in history, we are confronted nearly everywhere by mass opinion as the final determinant of political and economic action."

Back in 1888, Lord James Bryce called American public opinion "impalpable as the wind . . . , a judgment and sentiment . . . , which is imperfectly expressed . . . , is not to be measured . . . , is not easily gathered from the most diligent study." To this day, what we think the public believes often is only a dim, distorted, approximate reflection of reality that changes when examined from different angles. Despite the fact that in the last fifty years, the art, science, and mystique of measuring, analyzing, and manipulating public opinion have taken center stage, Lord Bryce's insight remains remarkably valid.

During the past several decades, a vast literature has emerged from an expanding industry of professional pollsters, marketers, and polit-

ical consultants, working hand in hand with a growing army of political scientists, sociologists, psychologists, communications experts,
economists, business consultants, and advertising, marketing, and consumer affairs specialists. Still, despite increasingly sophisticated efforts
to influence, control, and manipulate what people think, most political
scientists continue to regard public opinion with extreme wariness. By
and large, they don't trust the people at large to come to sound judgments about difficult issues, preferring to leave such decisions to experts and the professional elite. Fearful of too much democracy, they
view public opinion as a largely unpredictable, often irrational, and at
times untrustworthy force.

In this chapter we shall briefly explore the relatively new history of
the study of public opinion, examine the frequently questionable ways
in which the public's views are evaluated, reflect on public opinion's
changing role in political leadership, and look at research that de-
scribes how rational or irrational the American public's judgment has
been in the past five decades or so during which it has been studied
intensively. If, in the electronic republic, public opinion is indeed becoming the fourth branch of government, it is essential to have some
understanding of its nature.

We start with an episode from my own experience at PBS that demonstrates the fragility of public opinion, the unreliability and self-
serving nature of much of the data that purports to report what the
public thinks, and the need, even in the most democratic of environ-
ments, to take a position of leadership that at times may fly in the face
of perceived public sentiment. On May 12, 1980, PBS aired what was
probably its most controversial program ever, "Death of a Princess,"
a two-hour, made-for-television docudrama that started out as possibly the most unpopular program in PBS's history and ended up as the
highest-rated broadcast the network ever ran.

Docudramas happen to be my least favorite kind of program. Operating in the shadow between reality and make-believe, they are neither fact nor fiction. Under the pretense of authenticity and accuracy,
news docudramas take unacceptable license with the truth. During
what I regard as the low-water mark of network television news at
the end of the 1980s, the CBS and NBC news divisions briefly flirted
with the use of docudrama re-creations in their news broadcasts. Under pressure from callous new owners who cared more about profits
than integrity, the news divisions started to employ actors instead of
real news footage to heighten the drama and excitement of the stories
they were reporting. Fortunately, the fake news reports were trans-

parently phony. Neither the public nor the critics would tolerate them and the practice was quickly abandoned by both CBS and NBC News.

PBS, however, is not a news organization, and we carefully identified "Death of a Princess" on the air as the dramatization of a true story, although it was televised, much to my own discomfort, as part of the nonfiction *World* public affairs documentary series. A coproduction of public television station WGBH in Boston and a British commercial company, "Death of a Princess" reenacted the tragic story of a young Saudi princess and her lover, a commoner, both of whom had been executed in Saudi Arabia in 1977 for committing adultery: the princess was stoned to death; her betrothed beheaded. For obvious reasons, the program could not have been filmed as a legitimate news documentary. No film existed of the actual events or the principals in the story, and none could be photographed in Saudi Arabia. After months of investigation and surreptitious interviews with Saudis who had firsthand knowledge of the incident, actors were employed to reenact what presumably had actually happened.

For at least a month prior to its broadcast, "Death of a Princess" was at the vortex of a fast-spreading public firestorm. The Saudi royal family did everything possible to prevent PBS from broadcasting the film. Saudi spokesmen threatened dire economic and diplomatic reprisals against the United States if "Death of a Princess" were to air. The Saudis, the biggest oil supplier to the United States, let it be known that they were prepared to shut off the supply or raise oil prices precipitately. Energy experts, some recruited by the Saudis' American public relations firms and lobbyists, predicted that "Death of a Princess" would bring a severe oil shock in the United States. Saudi agents commissioned public opinion polls that, they said, showed a majority of Americans strongly opposed to broadcasting the film. And that majority supposedly grew as the broadcast date approached. Members of the public were quoted as saying that no television drama would be worth the economic disaster of a Saudi oil embargo in winter—not an unreasonable view. Survey after survey showed the public to be increasingly apprehensive about the broadcast.

The British had already experienced the Saudis' anger. A month before the PBS broadcast date, a commercial channel in Britain had televised "Death of a Princess"; the BBC—significantly less independent of government pressure than PBS—had earlier turned down the program. American newspapers were filled with ominous stories about the Saudis' angry retaliation against the British and the British government's unseemly efforts to placate their wrath. Foreign Secretary

Lord Carrington and his deputy sent their personal regrets to Saudi King Khalid for the offense "this particular television film had caused to the Royal Family in Saudi Arabia." Spurning the British apology, the Saudis expelled the British ambassador and canceled a long-planned visit by the British defense minister. With a good deal of publicity aimed at impressing the American public, the Saudis postponed granting landing rights for the British supersonic Concorde and canceled a state visit to Britain by the Saudi king. Members of the Saudi royal family were ordered home from London. Even worse, the Saudis announced they would review all their trade agreements with Britain and threatened to cancel major contracts with British companies. More than two billion dollars was reported to be at risk because of that one offending television program. Those actions were being carried out, it was clear, with an eye to pressuring PBS not to broadcast "Death of a Princess" in the United States.

As president of PBS, I had the responsibility to decide what to do. Four days before airtime, President Carter's acting Secretary of State, Warren M. Christopher, fired off a letter to me, which was simultaneously released to the press, urging that PBS "give appropriate consideration to the sensitive religious and cultural issues involved" in broadcasting "Death of a Princess." While Christopher wrote that "the government of the United States [did not seek] to exercise any power of censorship" over PBS, the import of his letter was clear—kill the show. It only added fuel to the fire.

Agents of the Saudis, I was surprised to learn, originally had drafted a far more threatening letter for the secretary to sign. They obviously assumed that the U.S. government could simply control the programming of PBS and order the broadcast killed. Christopher decided to take a somewhat less confrontational approach in keeping with Americans' skittishness about government interference with free speech, although his letter was unprecedented, as far as I know, in putting on paper the interest of the government of the United States in killing a television program. Enclosed with Christopher's letter, and also released to the press, was a tough message from the Saudi ambassador attacking the film and conveying his government's urgent request that "Death of a Princess" not be broadcast in this country.

In Congress, leaders of both parties weighed in to express concern about PBS's plan to run the film in the face of what they said was the public's clear opposition. Mobil Oil, which held a big stake in Saudi oil fields and is one of PBS's biggest and most visible underwriters, ran prominent newspaper ads urging us to heed public opinion and cancel the program. Mobil argued that the people endorsed that decision, "in the light of what is in the best interest of the U.S. Clearly the people of the United States have the right to expect that the media will not abuse its privilege." The Mobil ad concluded, "The public will have to decide whether a 'free press' is acting responsibly." Its logic, however, was dubious; if Mobil had its way, the public would have had to decide whether PBS was acting responsibly based on a film that nobody would have been allowed to see.

Admiral Thomas H. Moorer, retired chairman of the Joint Chiefs of Staff and member of the board of another prominent PBS underwriter, Texaco, phoned me from Dallas to say that public sentiment against the program was so strong that he would pay PBS not to run it. A group of Texas businessmen would put up the money to bury the broadcast, he said. Since the polls showed that most Americans did not want to see the program and since PBS was supported by public money, then PBS should follow the polls and heed public opinion, Admiral Moorer insisted.

A few PBS stations urged that we go ahead with the broadcast as planned, arguing that it was essential for the long-term integrity of public television to resist the pressure no matter what the public or others thought. Other stations demanded that we cancel the show in the national interest. The former governor of South Carolina, America's ambassador to Saudi Arabia at the time, urged PBS to heed the warnings and insisted that South Carolina, one of PBS's most influential state networks, refuse to broadcast "Death of a Princess."

Unlike most television scheduling debates, this one could not be settled quietly in the back room of the network's program department. As the clamor escalated, it became clear that the fight for public opinion had become the main battleground. Weighty matters of economics and foreign policy appeared to be at stake as were the integrity and independence of public television. But the perceived attitude of the people themselves was the key. The controversy grew so intense, the PBS board held an emergency meeting by telephone conference call. Was the public really up in arms? Were the surveys accurate? How deeply did people really care? As a matter of policy, the board did not get involved with individual program decisions, so the final determination was left up to me, although as Board Chairman Newton N. Minow pointed out sardonically, thanks to the Mobil ads and the Christopher letter, PBS could not retreat from the broadcast without seeming to compromise the integrity and independence of all of public television.

Never was there any question in my mind that the program should be broadcast. Despite my personal reservations about docudramas, it was certainly not an irresponsible program. And I didn't believe the reports about all the polls; they sounded too self-serving. Even if I were wrong about the polls, after all that publicity I decided that the nation's viewers should have the chance to see the program for themselves and make their own judgment about it.

"Death of a Princess" ran on PBS as scheduled, followed by a special live program that featured representatives of all sides of the great controversy discussing the issues. However negative public opinion may have been before the broadcast, after the broadcast it became abundantly clear that the public had turned totally supportive. With the help of all the publicity, "Death of a Princess" achieved a record-breaking public television audience. It seemed as if, after watching the show, viewers from one end of the country to the other simultaneously decided, "It's only a television show so what was all the fuss about?" The absence of negative reaction was almost eerie. PBS received applause and awards for standing up to the pressures and preserving its independence against threats from the U.S. government, the Saudis, and its own underwriters. An editorial in *The New York Times,* sharply critical of acting Secretary of State Christopher for "meddling" with the constitutional protection of the press, saluted us as the "protector of the Constitution."

Take a public opinion survey that asks people if they think a single TV program is worth the cost of a gasoline shortage in winter and a national economic crisis and their answer—sensibly—would be: "Absolutely not. Cancel it." Ask the very same people if PBS should kowtow to a foreign government's threats, or to American government arm-twisting, or to an oil underwriter's imperious demands, and the answer would be exactly the opposite: "Absolutely not. Run it." Any accurate reflection of public opinion, if it is going to be useful, must take into account not only what people think about a particular issue, assuming that opinion has crystallized, but also how deeply and intensely people feel about it. The vast majority may indeed express itself in favor of universal health care coverage and catastrophic health insurance, only to turn the other way as soon as their costs are revealed or they learn about the new taxes that will be necessary to pay for them.

THE CHANGING PERSPECTIVE ON
THE NATURE OF PUBLIC OPINION

Paying serious attention to public opinion is a relatively recent phenomenon. According to the Oxford English Dictionary, the term first
came into use in 1781. Its appearance coincided with the first major
stirring of democracy in Europe and America, and expressed the first
recognition that members of the public actually might be competent
to form judgments of their own. Interestingly, the use of the word
opinion rather than, say, a word like *judgment* in conjunction with
public, suggests a tentativeness, a lack of certainty, an idea in transition, something that is subject to change and perhaps in doubt. Kings,
emperors, and generals, by contrast, were never perceived as having
mere opinions or viewpoints; instead, they had certainty. They proclaimed decisions, which they handed down from above and which
were not to be called into question.

The early British and French democratic political philosophers, like
John Stuart Mill and Montesquieu, separated out the idea of the public
citizen, who was seen to be public-spirited and devoted to the common
good, from the role of the private person, who would selfishly pursue
his own self-interest. The rhetoric of the American Revolution was
among the earliest to portray ordinary citizens as thinking, reasoning,
and deliberative people who, therefore, were capable of governing
themselves and making judgments in the general interest. After the
American Revolution, as we have seen, a good deal of uncertainty
arose about this. The Founders feared the irrationality of public opinion and the Constitution sought to protect the new nation from its
excesses through a carefully contrived system of checks and balances.

After the French Revolution's Reign of Terror in 1793–94, the image of the public as a great beast took hold—irrational, intolerant,
emotional, a mob, the lowest form of common consciousness. The
public temper was perceived as capricious, like human nature itself or
the weather—a force to be lived with, accommodated when necessary,
and generally taken for granted. Today, the public and private spheres
are no longer considered separate and apart but are viewed as an
integrated whole, a mixture of both. And in the past fifty years, the
study of the complexities of public opinion has taken on extraordinary
momentum. The academic, business, and political communities have
joined forces in their efforts to understand, influence, entice—and market to—the people.

The seminal American examination of public opinion, Walter Lippmann's book, fittingly entitled *Public Opinion*, did not appear until 1922. As Lippmann wrote, "Since Public Opinion is supposed to be the prime mover in democracies, one might reasonably expect to find a vast literature. One does not find it. . . . On the sources from which . . . public opinions arise, on the processes by which they are derived, there is relatively little. The existence of a force called Public Opinion is in the main taken for granted."

Public opinion polling did not begin in earnest until the 1930s. The effort to understand public opinion—its theory, character, and origins—accelerated in a big way after World War II. Since that time, every nuance of American public opinion has been continuously scrutinized, measured, and analyzed. Both the technology and sampling theory have grown extraordinarily sophisticated. The original definition of public opinion, as the predominant attitude of a community or the collective will of the people, no longer suffices in view of today's increasingly intense efforts to understand and, if necessary, change people's attitudes. The public is divided into demographic, regional, racial, ethnic, and other interest groups. Computers enable pollsters to isolate and identify specific segments by every conceivable criterion—zip code, income, age, sex, race, religion, group affiliations, education, disease, and others even more obscure. Daily polls and surveys purport to measure people's sexual practices, eating habits, political preferences, television viewing, newspaper reading, child rearing, religious beliefs, leisure-time pursuits, and taste in clothes, cars, cigarettes, cartoons, and cosmetics. Such surveys are often dubious indicators of the public temper—one-dimensional snapshots, reflections of a reality that is at times fleeting and frequently barely considered.

Wide-area telephone dialing, automated call-back equipment, high speed processing, bar codes, people meters, and other technological marvels have turned the work of opinion surveying into an arcane, if sometimes pseudo-scientific and self-interested pursuit. Focus group interviews, telephone surveys, tracking studies, psychological profiles, street corner interviews, remote sensing devices, 900 numbers, and flash polls are staple tools used continuously in business as well as politics.

The politicians and their consultants have enthusiastically embraced the consumer industry's techniques of probing, analyzing, and quantifying their customers' attitudes and opinions. Politicians sell themselves and their political programs to citizens using the same marketing techniques as the packaged goods manufacturers use to sell products and services to consumers. Politics, like government itself, has expanded from a dedicated personal calling to a large, thriving, expensive service industry. The campaign managers, lobbyists, and media experts of the modern-day politics industry add their own special brand of political marketing and persuasion. In the interest group politics of today, each faction must first be understood, then accommodated, and if possible, brought together in coalitions.

The growing tendency to make the public part of the political decision-making equation can be costly and time-consuming and comes with certain dangers. The job of reaching and persuading the general public is invariably an extraordinarily expensive undertaking, usually affordable only by those interests with a good deal of money. Compromise, the essence of politics, becomes more difficult after one has staked out a position in public. To constituents who are deeply committed, a retreat can look like a betrayal. George Bush brought the "No more taxes, read my lips" issue front and center, and then was destroyed by it when he reversed course and compromised with the Democrats not long after his election in 1988.

Also, the more intensely the public feels about an issue, the higher the stakes become for those on the winning or losing side. When people care deeply about an issue like abortion, prayer in the schools, or health care reform, their disappointment at failure to get their way can produce a major political upheaval.

Finally, the more we learn about how to respond to and understand the public, the more we also increase the potential to influence, change, and even manipulate public opinion. In the electronic republic, political manipulation is the other side of the coin of effective political persuasion. What looks like manipulative propaganda to one, invariably is seen as an honest educational effort to another.

HOW DOES ANYONE KNOW WHAT THE PUBLIC ACTUALLY THINKS?

The increasing obeisance being paid to public opinion and its growing influence on the major decisions of state raise to a new level of concern the need to determine accurately what the public's views are on the issues that confront society. Apart from election day, how does anyone know for sure exactly what the general public thinks and what specifically people want their government to do at any given time? Elections have the virtue of producing clear-cut, definitive, and official results. They are scheduled at regular intervals, are conducted according to established legal procedures, are subject to official oversight, and have legal standing. Even though election results have been known to be fixed, sometimes even blatantly, severe criminal penalties are imposed for election fraud and deceit. Dishonest counting and ballot stuffing are not countenanced.

The increasing abundance of public opinion polls, on the other hand, operate entirely outside the framework of any official sanctions, regulations, or oversight. When it comes to polling, surveys, and public opinion research, caveat emptor is the rule, not the exception. Often the results are wrong, inadequate, untrustworthy, unreliable, and self-serving. And those are not problems that afflict politics only.

The highly sophisticated broadcasting industry offers an object lesson in the inadequacy of current standards of measuring public behavior. Although the entire multi-billion dollar financial structure of broadcasting depends on measuring its audiences—ratings determine how much advertisers pay—the industry's ratings data are notoriously inaccurate and unreliable. In the 1990s, the major television networks commissioned a comprehensive study—the Contam study—of Nielsen people-meter ratings, the mainstay of the entire television business. The study's findings were "absolutely devastating to the entire system," according to Dr. Ron Milavsky, former NBC executive in charge of audience measurement and later professor of research at the University of Connecticut. Referring to the findings, David Poltrack, CBS's senior marketing official, admitted, "The whole business is crazy. . . . I don't think there's an advertising agency in the United States that could get up in front of its clients and justify the way business is being done. It's being done on Nielsen ratings which probably aren't representative of the real population. . . . The way television and media are bought in this country is ridiculous."

To be sure, if information is wrong or misleading about which late-night host attracts the biggest audience or whether people prefer their soap packaged in a hard box or soft wrapper, the consequences to the nation are not terribly serious. At worst, flawed people-meter ratings or poll results may cancel a TV series that should have survived or cause a consumer goods manufacturer competitive pain. But if emotionally charged, life-and-death public issues, like war and peace, taxes, abortion, and gays in the military, are decided on the basis of what the president or the Congress thinks the people want, and if that information turns out to be wrong or misleading, the consequences could be devastating for the nation.

In the not too distant future, the United States may well decide that it has no choice but to introduce an official, legally recognized computerized system of electronic voting and public opinion polling to operate alongside the free market surveys. In the electronic republic, voting in federal plebiscites and on national initiatives and referenda may well become part of the legally sanctioned duties of citizenship. Different proposals along these lines have already been introduced by members of Congress from both parties.

Today, despite the important influence of public opinion polls on political behavior, skepticism abounds about their quality and integrity, and with good reason. For example, a poll conducted on the eve of America's entry into the Gulf War showed that 39 percent of Americans favored immediate entry into the war against Iraq. But as *The New York Times* later reported, "because that poll was conducted overnight, no effort was made to reach people who did not answer their phones the first time. Such a poll, survey experts say, may include disproportionate numbers of retired people and fewer young men, who are less often home than other groups." Another poll on the same topic, conducted over a four-day period, had more time to reach more people. It found that only 30 percent of the respondents favored going to war immediately—a significant 25 percent difference.

A poll commissioned by Ross Perot in March 1992 asked, "Should laws be passed to eliminate all possibilities of special interests giving huge sums of money to candidates?" It was what *The New York Times* called "a textbook case of how readily results of a survey can be manipulated by the pollster." Ninety-nine percent of those responding answered yes. Who would have answered that question any other way? When another polling company asked the question in a different form, "Do groups have the right to contribute to the candidate they support?" only 40 percent favored limits on contributions.

Studies that tried to verify even the most respected polling data found case after case in which those who are supposed to be sampled are not actually called, others who should be surveyed refuse to be interviewed, questions are biased, errors are made in interpreting responses, and the poor and hard to reach are severely underrepresented. An Advertising Research Foundation study of 7,000 survey interviews in 1975 discovered a 37 percent error rate. A study of 182 national surveys in 1978 found that 40 percent or more of selected respondents were not called and 28 percent of those who were called refused to be interviewed.

The credibility of polls that ask people to call a 900 telephone number with their reaction to a presidential speech or political issue is tainted by the inevitable self-selection of the respondents and the likely abuse by organized groups eager to overwhelm the available phone lines. In 1990, *USA Today* published the results of a poll that asked people to call a 900 number to say whether they thought Donald Trump represented "what was best in America or what was wrong with this country?" According to Dr. Stanley Presser, president of the American Association for Public Opinion Research, "Of 7,802 calls, 5,646 came from two phone numbers at a firm controlled by one of Trump's friends. He won in a landslide."

The reasons so much public opinion research is so bad are legion. Many polls are conducted not to find an honest answer but to reach a preconceived conclusion that is designed to bolster a special interest, as in the case of "Death of a Princess." Often, survey questions are biased, confusing, or sloppy. To save money, samples are not big enough to be meaningful. Pollsters are untrained and inexperienced. The public is too mobile to be counted accurately. Refusal rates keep escalating. The poor and minorities are inevitably undercounted. Results are incorrectly tabulated or misinterpreted, either accidentally or on purpose.

Most tabulations of public opinion settle for what sociologist Leo Bogart has described as "comic book statistics—flat, simplified treatment of the round, multi-layered, multi-faceted, intricate . . . real thing that is opinion." Law professor Lindsay Rogers had a more pithy description. The language of polls, he wrote, is "the language of baby talk—'yes,' 'no,' 'I don't know.' " Moreover, polls can lie. What people are willing to say out loud to a pollster may not reflect their private, politically unacceptable but actual opinion about a sensitive issue involving race, sexual preference, or ethnic or religious matters. A white voter, when asked if he or she would support an African-American for mayor, may say yes even though the voter has no intention of ever actually voting that way. The voter knows that an affirmative answer would be considered more respectable than the truth.

Finally, if the research itself is not flawed, often the "spin" that is put on the results is off-the-mark. A big percentage jump in the polls, coming from a very low base, can suggest a huge popular surge that in reality is quite meaningless, given the overall size of the vote.

Back in 1962, as a young and ambitious new director of advertising for NBC, I was involved in a memorable example of skewing survey results by interpreting them in a way calculated to serve my own interests. When I first came to work for the network, I was confronted with a nasty dilemma. NBC had commissioned an elaborate and expensive audience survey that seemed to prove definitively that newspaper tune-in advertising did not work. At best the ads would attract a very small incremental audience to the network. And the study found that the ads cost more than the network could earn from advertisers for the additional viewers the newspaper ads brought in. Based on the findings of that study, NBC canceled all of its newspaper advertising to promote the upcoming television season. Having no advertising budget was no way to launch a career as the network's new advertising director, eager to prove how good he could be. So I conspired with a bright young NBC researcher named Paul Klein (who years later rose to become the network's head of programming), to reinterpret the survey results to show that newspaper tune-in advertising worked. We argued that in the three-network audience race, the last quarter of a rating point was the key to winning the audience crown. The network that could boost its audience by that tiny incremental rating percentage would be able to charge its sponsors golden rates for its entire program schedule. The ads, expensive though they were, could put NBC over the top.

To the dismay of the network's chief administrative officer, who had commissioned the study in the first place in the hope of saving money, we persuaded NBC's demoniacally competitive president, Robert E. Kintner, not only to restore the network's newspaper advertising budget but to add to it. The same survey findings that were first used to kill the ad budget were reinterpreted to justify a fat increase. To this day I do not know which interpretation was more legitimate. I simply followed the customary industry practice of using whatever statistical "spin" best served my own interest. It is what political consultants refer to in another context as "true lies."

Several years later, Paul Klein and I again joined forces to urge that NBC convert its entire program schedule to color on the grounds that, despite the added cost, every new color television set would be adjusted to the NBC channel, giving the network an inherent ratings advantage over its competitors. Of course, it didn't hurt that NBC's owner, RCA, was the big manufacturer of color television sets.

In a democratic system increasingly dependent on public opinion and public response, the ways to measure that opinion and judge that response remain dangerously untrustworthy and susceptible to self-serving manipulation. In the electronic republic, the art of understanding and accurately interpreting public opinion will be a skill even more demanding and highly prized than it is today.

THE CHANGING ROLE OF LEADERSHIP IN THE ELECTRONIC REPUBLIC

What does the increasing role of ordinary citizens in the day-to-day decisions of government do to the nature of political leadership? Since the dawn of democracy in ancient Greece, philosophers and politicians—observers and participants alike—have debated the nature of political leadership: Should democratic leaders lead or follow the public will? Most of the time, that debate was carried on without anyone having a very good idea of what the public will was about most issues, except at election time when the ballots were counted. Since World War II, that changed. Today, leaders have to make decisions in the light of what the polls tell them the public is thinking on any given day, sometimes at any given hour.

History admires strong, bold, intuitive leaders. It is less kind to those who "keep their ear to the ground," which, as British Prime Minister Winston Churchill suggested, is hardly a position that encourages citizens to look up to their leaders. "Nothing is more dangerous," Churchill said, "than to live in the temperamental atmosphere of a Gallup Poll, always feeling one's pulse and taking one's temperature." In a similar vein, a generation earlier, Lord Bryce wrote, "The duty . . . of a patriotic statesman in a country where public opinion rules would seem to be rather to resist and correct than to encourage the dominant sentiment. . . . He will confront it, lecture it, remind it that it is fallible, rouse it out of its complacency."

We are accustomed to hearing leaders insist that their job is to do what is right, not what is popular. Asked at a Rose Garden press conference why his approval ratings had taken a nose dive, President Bill Clinton responded, "If I worried about the poll ratings, I'd never get anything done here. . . . My job is to do my job and let the chips fall where they may." Such bold words are calculated to put a brave face on the president's declining popularity, even if Clinton, like all other recent presidents, actually does not make a major move without consulting his own full-time resident pollster.

The classic description of the role of the representative leader in a democracy was given in 1774 by one of Great Britain's most distinguished parliamentarians, Sir Edmund Burke, in an often-quoted speech to the Electors of Bristol: "Your representative owes you, not his industry only, but his judgment; and he betrays, instead of serving you, if he sacrifices it to your opinion. If government were a matter of will upon any side, yours, without question, ought to be superior. But government and legislation are matters of reason and judgment, and not of inclination; and what sort of reason is that, in which the determination precedes the discussion; in which one set of men deliberate, and another decide. . . . I know that you chose me . . . to be a pillar of the state and not a weathercock on top of the edifice . . . , of no use but to indicate the shiftings of every fashionable gale."

A noble sentiment, but spoken long before the days of instantaneous interactive telecommunications when he could have learned hour-by-hour what the Electors of Bristol were thinking. Ironically, Lord Burke suffered the fate of many elected leaders who fail to heed "the shiftings" of the popular gale. During his first Parliamentary term, he championed laws that turned out to be unpopular with his Bristol constituents. Fearful of losing the next election, Burke changed districts and stood for Parliament from a safe "rotten borough," where he no longer would have real constituents to please. So much for his brave words about the nature of political leadership.

In the new era of instantaneous and ubiquitous electronic communications, it has become increasingly apparent that leaders will lead effectively only if they can bring the people along with them. In the United States, government has become a permanent political campaign. Public opinion must be courted and counted (although not always followed) before anything important can be undertaken. Nothing gets done in Washington without first trying to figure out how the public will react. Strong leadership is no longer a matter of pronouncing God's will to the public in an intuitive and paternalistic way. Today, the strong leader must also listen attentively and respond effectively to people's wishes; good followership has become almost as important as good leadership.

HOW RATIONAL IS
PUBLIC OPINION?

What, therefore, can be said about the quality of the American people's judgment? Recent literature on public opinion has managed to shed fascinating new light on that age-old question.

Hamilton complained back in 1787 that "The voice of the people has been said to be the voice of God and however generally this maxim has been quoted and believed, it is not true in fact. The people are turbulent and changing; they seldom judge and determine right." And consider Lippmann's argument that "The general tendency to please the electorate . . . leads to insolvency, towards the insecurity of factionalism, towards erosion of liberty and towards hyperbolic wars. The pressure of the electorate is towards soft rather than hard decisions." According to Lippmann, "That a plurality of the people sampled in a poll think one way has no bearing upon whether it is sound public policy. . . . The statistical sum of their opinions is not the final verdict on an issue. It is rather the beginning of the argument." If Hamilton, Lippmann, and so many others are vindicated in their view that public opinion "seldom judge[s] and determine[s] right," then an electronic republic that increasingly has to rely on the people's judgment faces an exceptionally bleak future.

In an effort to seek definitive evidence about the competence of public opinion in the United States, two professors, Benjamin I. Page of the University of Illinois and Robert Y. Shapiro of Columbia University, conducted an exhaustive analysis of Americans' national policy preferences by examining the data from every published poll since the practice of polling began a little over half a century ago. Their massive study, *The Rational Public*, published in 1992, was compiled from what the authors say is the largest, most comprehensive set of information about public opinion ever assembled, literally thousands of national polls taken in the United States between 1935 and 1990.

As the book's title suggests, the professors found that the American public tends to make sensible, coherent, and consistent judgments. They concluded that people as a whole react to events and to social and economic changes in reasonable and predictable ways. While the views of individuals can be, and often are, "ignorant, shallow and unstable," Page and Shapiro found that the evidence shows "public opinion as a collective phenomenon is stable, meaningful, rational and able to distinguish between good and bad."

Page and Shapiro explained the apparent contradiction between the relative inadequacy of so many individually held opinions and the collective phenomenon of a "rational public" largely on the basis of what is known as the "statistical aggregation process." More than two centuries ago, the French philosopher Condorcet demonstrated mathematically that a collective decision by majority vote will have a much higher probability of being correct than any individual answering alone. In a large group, individual aberrations, ignorance, and other inadequacies tend to cancel each other out. (The same logic explains why a jury, as a collective body, is more likely to come to a sound and reasoned judgment than a judge or juror deciding alone, no matter how smart or well-informed the judge or juror might be.) A majority vote among many citizens will have a very high chance of correctly reflecting the general interest compared to any individual voting by him or herself. Of course, the real world, especially the real political world, is filled with false information, deceptive arguments, exaggerated claims, unscrupulous advocates, seductive propaganda, and manipulative techniques of persuasion. The popular majority can go off the track of reason, at times with appalling consequences. American history is filled with such examples: the Salem witchcraft trials in the seventeenth century, the hunts for Socialists and Reds in the early twentieth century, the incarceration of Americans of Japanese extraction during World War II, the McCarthy Red-baiting era early in the cold war.

In each of those instances, however, the political leaders and elite of the time, the very people whom Madison and Hamilton counted on to rise above mob tyranny and diffuse mass hysteria, were the ones most responsible for setting and fanning the flames. Certainly, the people are not always right. Nobody is always right. But a convincing case can be made that when the people have been terribly wrong, the problem usually had a good deal more to do with bad leadership and inadequate or misleading information from the top than irrational pressure from below.

Page and Shapiro were not naively suggesting that public opinion is "the voice of God," and they certainly did not argue that it has been infallible. They simply found that the public's collective judgment tends to be better than the sum of its individual parts. The views of the public at large are usually no more irrational or contrary to the public interest than the individual opinions of the so-called governing elite. This is true despite the fact that the nation's political leaders invariably are better informed, have more experience, and are able to spend far more time deliberating over any given issue than can those they represent. The nation's governing officials have the constitutional responsibility to exercise cool reflection and shield the nation from the tyranny of ill-informed or emotionally aroused majorities. But there is no evidence that they have exercised that responsibility when it has been most needed.

Page and Shapiro's definitive and reassuring analysis of national public opinion data accumulated over half a century, confirms the findings of a good many other careful, if less ambitious studies of voter rationality that have examined the results of hundreds of state referenda and initiatives. The state referenda and initiatives are the closest the country comes to practicing direct democracy, and the studies of their results have come to essentially the same conclusion as Page and Shapiro in *The Rational Public*: People tend to have reasoned and sensible views about public issues and therefore, when asked to make political decisions, they tend to do so reasonably and sensibly.

A comprehensive study of the response of Oregon voters to initiatives in the 1950s found that "over the long period, the electorate is not likely to do anything more foolish than the legislature is likely to do. Both have bouts of wisdom and both have erred. The risk that people will enact something catastrophic is not great."

In 1980, political scientist James A. Meader examined voter competence in South Dakota based on the results of the state's initiatives and concluded, "The voters are capable of taking a long range outlook when they consider initiatives on the ballot. Rather than opt for a short range financial benefit, the voters showed a stronger concern for maintaining . . . [what] will enhance quality of life in South Dakota into the future." Meader found that by any standard, even if one disagrees with their decisions, voters exercised informed judgment and rational choices.

That same year, a New York state senate committee, looking into the proposal to adopt the use of initiatives and referenda in New York, studied voter performance on initiatives in all of the states that already had them. The committee concluded that "voters were able to look beyond self-interest and take into account what was good for state, government and society, and appreciated the opportunity to participate."

A study of 115 referenda in California by Dr. Max Radin of the University of California concluded, "One thing is clear. The vote of the people is eminently sane. The danger apprehended that quack nostrums in public policy can be forced upon the voters by demagogues is demonstrably nonexistent. The representative legislature is much more susceptible to such influences."

Of course, money plays an increasingly influential role in persuading the public and derailing public opinion, just as it has always had a major influence on the decisions of many, if not most, public officials. As we shall discuss in more detail later, numerous studies have found a significant correlation between the amount of money spent to influence voters and the outcome of popular initiatives and referenda. In 1976, political scientist John S. Shockley found that while so-called big money has only about a 25 percent success rate in promoting ballot issues, it has a 75 percent better chance of defeating ballot issues than the less wealthy side. A study of fifty major ballot questions in four states (California, Massachusetts, Minnesota, and Oregon) between 1976 and 1982, concluded that "campaign spending is the single most powerful predictor of who wins and who loses," a fact that is equally true of legislative battles.

The most interesting general conclusion to be drawn from all of these studies is that, at the very least, public judgments have tended to be no worse than the judgments of the political elite. American public opinion, an increasingly powerful and influential force in the electronic republic, has been found over the years to be, in general, reassuringly sensible and competent.

4

THE AGE OF
MEDIA POWER

No discussion of American politics and certainly no discussion of the electronic republic can ignore the role of the media, old and new, print and electronic. The media link the government to its more than two hundred million citizens, and link those citizens to the world outside. The impact of the media on politics, not to speak of their impact on all of society, is so pervasive and so all-embracing that it is not easy to sort out the media's many roles and their specific effect on our lives.

The Press, as its owners are fond of pointing out when lobbying against a new government restriction or for a special exemption, is the only industry specifically mentioned in the U.S. Constitution and given special constitutional protection. Under the First Amendment, "Congress shall make no law . . . abridging the freedom of speech, or of the Press" (with a capital P). The Press (or the Media, as it has come to be known in the age of electronic communications), has capitalized on its privileged position to achieve primordial power, influence, and wealth.

"It sets the agenda for public discussion and this sweeping power is unrestrained by any law," wrote journalist Theodore White in 1972, referring to the press. "It determines what people think about and write about, an authority that, in other nations, is reserved for tyrants, priests, parties and mandarins." Political scientist Walter Dean Burnham called "the dominance of the media over our politics . . . a monstrosity that presents a grave danger to what is left of democracy in the United States." Austin Ranney of the American Enterprise Institute described the media as "the new elector of the modern political age . . . , the opposing party, the shadow cabinet."

We have no shortage of descriptions of the media's awesome clout. The dean of the Columbia University School of Journalism informed incoming students of their special responsibilities in their newly chosen calling: "This is the age of media power. We set the agenda. We are the carriers of the culture and its values. . . . We are the brokers of information and ideas. Our decisions, our news judgments tell the people who they are, what they are doing, what's important and what they need to know."

Notwithstanding the special place of the press in the Constitution, it has not *always* been the thousand-pound gorilla of American politics, the central nervous system of American society, or the glue that holds the country together. In the early days of the republic the press was hardly an important political player. Newspapers served largely as partisan promotional vehicles for political factions, and personal propaganda outlets for political figures. Publications that achieved large circulation and widespread political influence, such as Paine's *Common Sense,* were rare. Politics relied mainly on the spoken word, communicated to the people directly and personally, unamplified, in face-to-face meetings, assemblies, discussions, and debates, often with the entire community in attendance. Politics was the general public's chief leisure-time activity. Only in this century has politics become virtually confined to, and channeled through, the media.

The unambiguous words of the First Amendment, adopted in 1791, may have ordered Congress to keep its hands off the press, but when Congress had established the State Department in 1789, it authorized the secretary of state to designate at least three newspapers as official printers of new laws and government proceedings. Far from being an independent institution, the political press in those days was a kept press, bought and paid for by the government, the politicians, and their parties. The examples are legion: The nation's first secretary of state, Thomas Jefferson, wasted no time putting on the State Department's payroll a Princeton classmate of fellow Virginian James Madison, Philip Freneau, who was hired to start a new newspaper, the *National Gazette.* Freneau's paper served as the mouthpiece of Jefferson's Republican Party. His job was to respond to the barrage of excoriating personal attacks published by the *Gazette of the United States,* a Federalist Party paper that lived off government advertising from Alexander Hamilton's Treasury Department. In New Hampshire, the editor of the *New Hampshire Patriot,* Isaac Hill, used his paper to build a party machine that carried the state for John Quincy Adams in 1828. In Virginia, the editor of the *Richmond Enquirer,* Thomas Ritchie, started his paper to promote Jefferson's and Madison's anti-Federalist Republican Party. In Washington, D.C., Duff Green's *Telegraph* served as the official voice of the Andrew Jackson administration. When Jackson grew disenchanted with Green, he simply shifted the government's ads from the *Telegraph* to a new journal, the *Globe,* edited by his friend Francis Preston Blair. As historian Daniel Boorstin observed, "Everywhere [newspapers] owed their establishment to government subsidy. The bulk of what issued from the presses was government work."

In those days, nobody ever thought of the press as politically independent or objective. Newspapers existed primarily to promote political causes. Printed on hand presses, they had small circulations, and were essentially house organs for political factions. In 1800, the average circulation for dailies, weeklies, and semiweeklies was seven hundred copies each. By 1830 the country had 650 weeklies and 65 dailies with an average circulation of twelve hundred copies. Their contents, usually filled with classical references, literary allusions, and scholarly rhetoric, were hardly suited for consumption by the average American citizen. Consider this typical bashing of Jefferson by Hamilton's *Gazette of the United States:* Jefferson's naked ambition would be uncovered, the *Gazette* wrote, "when the visor of Stoicism is plucked from the brow of the epicurean; when the plain garb of Quaker simplicity is stripped from the concealed voluptuary; when Caesar coyly refusing the proffered diadem, is seen to be Caesar rejecting the trappings by grasping the substance of imperial domination."

In the towns and villages of early America, politics was personal, social, and almost entirely oral. Political forums, mass rallies, speeches, torchlight parades, bonfires, meetings, debates, and discussions lasted for hours and involved entire communities. Families turned out en masse to listen to the orations and participate in the fun. In the absence of such modern-day distractions as professional sports, movies, theater, or television, politics served as the chief leisure-time entertainment. The written word carried little political influence in the late eighteenth century and early nineteenth century, at least compared to the spoken word, and was rarely a significant factor in the life of the general public. (Although it is fascinating today to read the nearly two-hundred-year-old criticism of sound-bite journalism in a letter from a Federalist leader to a friend complaining about the unseemly new Republican Party tactics of using the press to manipulate the electorate for political purposes: "Much was made of slogans and catch-words. . . . Intricate issues were summarized in a few phrases or re-duced to a line of wretched verse. . . . Political problems which might have stymied Solomon were resolved in a pun or an epigram. . . . Editors ran the same cant words and phrases in their columns over and over again . . . to rely on the effect of repetition.")

THE RISE OF THE PENNY PRESS

In the 1830s, the introduction of the steam-driven printing press not only altered the newspaper landscape but also produced major changes in the way politics was conducted. Dramatic technological advances in typesetting and printing lowered the price of daily newspapers, which—even at a penny a copy—grew into substantial businesses in the major cities. No longer mere house organs for competing political factions, newspapers became influential and independent institutions in their own right. Their circulation growth was phenomenal. By 1840, the country had 1,141 weeklies and 138 dailies with an average cir-culation of twenty-two hundred copies (for a total daily circulation of approximately three hundred thousand). By the end of the Jackson presidency in 1837, the *New York Sun* alone had a circulation of fifteen thousand. The United States became the greatest newspaper-reading country in the world. The popular penny press displaced the small circulation partisan press as the model of the daily newspaper.

The penny papers invented the idea of news. Whereas early news-papers offered verbatim reports of trials, presidential addresses, and letters from congressmen, the penny press broadened their coverage to include local crime, police incidents, social events, and the like. By comparison to the first newspapers, the penny press was actually rather indifferent to politics. Other than what was immediately local, news in the early to mid-nineteenth century traveled too slowly for the daily papers to report. Word of George Washington's death took five days to reach New York. News of the end of the War of 1812 took months to travel around the country, which is why the Battle of New Orleans was fought long after the war had ended. Reports of the battle of Gettysburg took a year or more to reach the outlying frontier territories.

In quest of mass circulation and advertising support, the major city newspapers gradually developed a tradition of political and journalistic independence. Their patrons changed from politicians and parties to advertisers and readers. Government subsidies in the form of legal notices shrank while circulation and advertising income rose dramatically. In the late 1820s, newspapers began to use paid reporters for the first time. The language of the new press went down-market, using words and anecdotes that the general public could easily grasp. Their appeal cut across party lines.

While the party press had dominated American journalism for the first part of the nineteenth century, as circulation runs increased the picture changed radically. Partisan political messages were confined mostly, although not entirely, to the editorial page and signed opinion columns. The large circulation penny press of the 1840s, like the new media of the 1990s, was derided as "sensational," "cheap," and "irresponsible" by the traditional six-penny press.

Popularization of news was accelerated in the 1840s with the introduction of telegraph wire services. Samuel F. B. Morse received his patent for the telegraph in 1840. By 1844, a Washington newspaper started printing telegraphed news from Maryland. The Utica, N.Y., *Gazette* hired an Albany correspondent to report from the state capital by telegraph in 1846 and within three months, ten upstate New York newspapers shared news by telegraph. In 1848, to take greater advantage of the telegraph wire's speed in transmitting the news, a group of New York newspapers organized the Associated Press.

Because telegraph systems charged by the word, reporting in spare, factual, unadorned prose was encouraged. The telegraph also brought a sense of timeliness to daily newspapers. And since the Associated Press served "papers with widely different political allegiances, it could succeed only by making its reporting 'objective' enough to be acceptable to all of its members and clients." The news wires delivered timely reports over great distances and served a fast-increasing number of papers in many towns and cities and across many state borders. The more uniform, concise, and anonymous the prose they transmitted, the more useful their information became to more newspaper clients.

As one press historian noted, "The wire services demanded language stripped of the local, the regional, and the colloquial. . . . If a story was to be understood in the same way from Maine to California, language had to be flattened out and standardized. The telegraph, therefore, led to the disappearance of . . . styles of journalism and storytelling. . . . The origins of objectivity, then, lie in the necessity of stretching language in space over the long lines of Western Union."

The electronic technology of the telegraph, followed by the under-seas international cable, caused vast changes in the language and focus of journalism. But apart from technology, the great changes in American newspapers after the 1830s were caused by "the changeover from the political culture of gentry rule to the ideal and institutional fact of mass democracy." The Jackson era, from 1829 to 1837 and extending into the Martin Van Buren administration that followed, brought populism to the fore. The Jacksonian period broadened male suffrage, put new emphasis on and gave new legitimacy to the common man, and stressed political equality and equal economic opportunity. It also brought on the cheap press, public schooling, and the rise of an egalitarian marketplace democracy where "money had new power and pursuit of personal self-interest had new honor."

One influence of the telegraph on newspapers was to substitute a new breed of reporter—the stringer who wired the bare facts—for the occasional correspondent who had written and posted leisurely, discursive, personal essays from distant places describing events in detail and interpreting their significance. Based on the stringers' barebones filings, stories were written in a rush in the newspaper office on the other end of the telegraphic line by reporters who never left their desks and never went out to report. The story became divorced from the storyteller and grew increasingly remote from the reader's local and personal experience. Both government and politics grew more distant and removed. Filtered through the printed word of newspaper pages, politics became less personal and more distant from home. The direct experience of oral communication was displaced by the second-hand experience of the written word.

This process was accelerated in the 1860s during the Civil War. Newspapers dispatched dozens of reporters to the front to send back eyewitness accounts of the battles, which were read avidly by the wives and parents in the cities back home. But the true era of reporters did not arrive until the Spanish-American War in 1898. In December 1896, William Randolph Hearst, who had recently bought the *New York Journal,* dispatched popular writer Richard Harding Davis and artist Frederic Remington to Cuba to cover the Cuban insurrection against Spain. Frustrated at being barred from the war zone, Remington wired Hearst, "Everything is quiet. There is no trouble here. There will be no war. Wish to return." Hearst is said to have replied with his now famous message, "Please stay. You furnish the pictures and

I'll furnish the war." The press had become a prime mover in determining government policy and influencing public opinion.

As the twentieth century matured, newspapers grew more corporate and less engaged with their own communities. Most big city newspapers, bought up by chains, were operated from out of town as bottom-line corporate businesses. Newspapers came to be treated by their chain owners as business properties like any other, whose primary goal is to maximize income for the shareholders. The colorful, idiosyncratic, headstrong press barons—the Hearsts, Pulitzers, Pattersons, Chandlers and McCormicks—who originally had put their own highly personal stamp on their papers, died and were succeeded by bottom-line-driven corporate managers.

At the turn of the twentieth century, New York *Daily News* publisher Frank Munsey was typical of the new breed. A success in magazines, Munsey bought the *News* in 1905. In his newspaper publishing career, Munsey succeeded in killing off eight or more newspapers. When he died in 1925, the legendary Kansas newspaper editor William Allen White wrote, "Frank A. Munsey contributed to the journalism of his day the talent of a meat-packer, the morals of a money-changer, and the manners of an undertaker. He and his kind have about succeeded in transforming a once noble profession into an eight percent security."

The corporate culture came to dominate the news business, treating news as a commodity or service no different from "toasters, light bulbs, or jet engines," to quote John F. Welch, chairman of General Electric, which bought NBC in 1986. Welch insisted that NBC News had no greater responsibility for public service than any of GE's more traditional lines of business, regardless of news' special Constitutional standing and the broadcast company's historic FCC license obligations. For Welch and a good many of his GE colleagues, news—even though part of hugely profitable NBC—was expected to make the same profit margins as every other GE division. As soon became evident, they had no qualms about doing whatever was necessary to achieve that goal, with little regard for journalistic standards, integrity, or taste.

As the mass media gradually became the main connecting links between the public and politics, the mass rallies, community debates, and political party picnics slowly disappeared. The media, along with other popular diversions and entertainments, including the automobile, professional sports events, movies, and shopping malls took their place. Politics no longer served as the centerpiece of citizens' public lives. Personal participation in American politics and political parties declined precipitately and changed its character and focus. That trend had started as far back as the end of the nineteenth century. Commenting on the 1892 election campaign, the *New York Herald* noted with approval "an unprecedented absence of noisy demonstrations, popular excitement and that high pressure enthusiasm which used to find vent in brass bands, drum and trumpet fanfarenade, boisterous parades by day and torchlight processionals by night, vociferous hurrahs, campaign songs, barbecues and what not." It was, the *Herald* concluded, "the dawn of a new era in American politics." Today, we look back nostalgically to the good old days of intense popular participation in the sociable and noisy excitement of politics.

THE RISE OF RADIO AND TELEVISION

The new era had dawned even before the growth of television—the displacement of print, which had always been local, by electronics, which essentially has no borders; the substitution of the sensory experience of sound and pictures for the deliberative and thoughtful written word. The revolution began with the invention of motion picture film early in the twentieth century. The shaky black-and-white newsreels of World War I gave ordinary people a new way to see for themselves far-off happenings and distant reality. The electronic communications revolution exploded in the 1920s with the emergence of radio. And a quarter of a century later came the triumphant debut and unprecedented expansion of television, which went from seven thousand households in 1946, to nineteen million in 1952, to forty-five million in 1960, to virtually every household in the country today. Television, the focus of the next chapter as well as this one, proceeded to change everything.

Newspaper readership has declined to the point where all but a handful of communities in the United States are now one-newspaper towns. In 1916, the number of dailies was at its peak, approximately twenty-five hundred. Today, the total has fallen below sixteen hundred. In 1910, only sixty-two dailies were owned by chains, which averaged less than five newspapers per chain. Today, more than 75 percent of all dailies are chain-owned, many owning dozens or more. In 1960, 80 percent of all U.S. adults read at least one daily newspaper. The figure today, according to a Simmons study for the National

Newspaper Association, is approximately 60 percent. The most powerful and influential newspapers left today are not primarily local but national—the *Wall Street Journal, The New York Times,* and *USA Today.* Others, like the *Washington Post, Chicago Tribune,* and *Los Angeles Times,* retain great influence in their respective communities even if they are gradually losing circulation.

A worse fate has befallen the general interest, mass circulation magazines, once the dominant national media. Seemingly indestructible when I started my first full-time job in the circulation promotion department of *Look* in 1954, general interest magazines were wiped out within a decade. *Look, Life, Collier's, Saturday Evening Post*—all were devoured by the voracious new medium of television.

By any measure, television has emerged as the dominant public force of our time, the epicenter of our existence. The television set in the average home is on more than seven hours a day. It fills more than half of the average American's free time—more than four hours a day, every day of the week. In less than half a century, television has transformed every major American institution and all of American society—the family, religion, work, leisure, education, politics, sports, the arts.

A fascinating and infuriating, at times baffling and contradictory combination of good and evil, reality and fantasy, quality and kitsch, importance and triviality, television is at one and the same time frivolous, superficial, and also central to our lives. It continues to change our society in ways we do not fully understand and, like children growing up before our eyes, may not even notice until the day we realize that so much has become so very different and that television clearly had a great deal to do with it.

Television is the ultimate fun-house mirror. It reflects, distorts, and exaggerates the best and the worst of who we are and what we are. When it comes to politics, and especially to elections, television is everyone's favorite whipping boy. Politicians, voters, political scientists, and even some of television's own correspondents level withering criticisms of its performance and of the appalling lack of public responsibility by most of its owners. Many deplore the medium's relentless intrusions into the private lives of public people. Critics focus on its shallow and misleading sound-bite coverage; its unrelieved negativism; its deteriorating standards of integrity; its mindless sensationalism; its failure to deal seriously with critical political, social, and economic problems; its disproportionate focus on crises and confrontation; its simpleminded preoccupation with polls and the "horse race" during political campaigns; its "inside baseball" reporting of political tactics rather than substance; its "feeding frenzy" mentality; its marketplace rather than public service orientation; its constant, carping criticism, and paradoxically, its increasingly close relationship with the governing establishment.

Television has become the irresistible villain of choice, blamed for almost all of society's man-made ills: "amusing ourselves to death," depriving us of "our sense of place," turning us into "couch potatoes," destroying our individuality, corrupting our moral values, saturating us with sex and violence, glamorizing crime, debasing the political process, creating a nation of insatiable consumers, ignoring our responsibilities as citizens, diminishing our sense of community, warping our perspective of the world, vulgarizing our arts, coarsening our culture, shortening our attention span, eroding our educational standards, undermining our religious faith, and distracting us from pressing social and economic issues—"the plug-in drug."

Complaints against television's role in the political sphere center on its expanding use of negative attack advertising, the dominant and corrupting role of big money in political life, the cynical manipulation of news by the politicians' media managers and spin doctors, and the avoidance of serious, thoughtful, and honest discussion of vital issues. To the general public, as well as to political scientists, the media—especially television—share with the politicians equal responsibility for the low state of the nation's public realm.

Buried somewhere in all that hyperbole is a good deal of truth. The mainstream news media share two essential characteristics that fundamentally affect the public's view of reality and its perspective on the American political process. Newspapers, newsmagazines, radio, and television are perceived to have certain real or imagined biases in their attitude toward the world and their approach to the news. And the very timeliness and episodic quality that define the nature of news, its constant "newness," if you will, give people a disconnected, and, in the end, superficial crisis view of reality.

THE BIAS OF THE MASS MEDIA

The mainstream mass media are an integral part of the national and international corporate community, dominated by modern-day corporate culture and driven by bottom-line financial considerations and marketplace demands. Yet, the most frequent present day charge against the mainstream media still centers on their liberal bias. They are viewed as being "an agent of social change" that "fosters populist suspicions of traditional mores and institutions." No matter what the political outlook of the media's corporate owners and managers, most newsmen and women, it is argued, tilt to the left, vote for Democrats over Republicans, come down on the side of progressive rather than conservative politics, and are generally nasty to business and to whatever administration is in power.

That view was given credence by two well-publicized studies, one by academics Stanley Rothman and Robert and Linda Lichter; the other by political scientist Michael Robinson. The Rothman-Lichter study interviewed 240 journalists and news executives at major news organizations and found that in the four elections from 1964 to 1976 those interviewed had voted for Democratic presidents by a margin of more than five to one. The study also reported that most of the major news organization journalists favored so-called liberal positions on abortion, homosexuality, and foreign aid. The Robinson study found that during the 1984 presidential campaign, the three television networks gave decidedly more negative coverage to Ronald Reagan and George Bush than to Walter Mondale and Geraldine Ferraro.

Sociologist Herbert J. Gans, while critical of the Rothman-Lichter study on the grounds that a reporter's vote one way does not necessarily mean reporting skewed in that direction, found that the news people he surveyed in the late 1970s reflected a kind of "early 20th century progressivism" in their essentially liberal values. Gans's conclusion reinforced the findings of Edward J. Epstein in his own influential observation of how network news works, *News from Nowhere*, published in 1973.

Since those studies, however, journalists and columnists for the national mainstream news media—whose work sets the tone for what is seen, heard, and read—have grown considerably richer and more deeply ensconced in the establishment world they cover. (On smaller newspapers, however, starting wages have fallen substantially since 1990.) In the big cities, the best known anchors and correspondents, not only in television but also in print, have become stars, commanding huge salaries, stock options, lecture fees, and other financial benefits in keeping with their celebrity status. They also command retinues of agents, lawyers, publicists, investment counselors, and flunkies. No longer do they resemble the breed of hard drinking, cynical, streetwise, faintly disreputable, working class Chicago reporters who were immortalized in Ben Hecht's and Charles MacArthur's drama *Front*

Page. Instead, the well-educated, graduate school alumni who inhabit the newspapers and newsrooms of today are upper-middle-class professionals who settle in the suburbs, play golf and tennis, and socialize with corporate executives, business leaders, government officials, and doctors and lawyers. In lifestyle and behavior today's mainstream journalists are part of the upper class or at least upper-middle-class establishment.

How do we reconcile the new breed with the adversarial, hypercritical culture of the news? My own experience suggests that journalists seem pro-change because change is news and status quo is not. They thrive in an atmosphere of change, crisis, and conflict. Their intense distrust of government and cynicism about the politicians they cover reflect the rise of the critical culture since the 1960s that characterizes much of American society. It is not a left-wing or liberal bias, but a bias toward what critic Lionel Trilling called the "adversary culture" that characterizes modern journalism. Distrust of government that rose during the opposition to the war in Vietnam has spread widely. "In 1958, 24 percent of the population felt 'you cannot trust the government to do right,' while 57 percent felt that way in 1973; in 1958 18 percent felt government was run for the benefit of the few, while 67 percent felt that way in 1973." In the two decades since, those negative feelings have grown even more intense and widespread. And among those who cover the news, who see up close just how government actually works, how much it tries to hide and deceive, and how frequently special interests get their way, the sense of cynicism and antagonism toward the political system has only been magnified.

Those who work for the mainstream media are citizens, too, subject to the same views and currents of belief as everyone else. And, as one observer remarked, "Journalists, especially those covering national politics, [are] . . . more deeply affected than most citizens because they . . . trusted more in, and cared more about, government." My own experience confirms this insight.

In 1969, after a series of enormously destructive race riots tore apart most of the nation's biggest cities, I organized an integrated group of television professionals, journalists, community leaders, and financial investors in the New York metropolitan area to challenge the license of one of New York City's oldest and worst commercial television stations. Our group was called Forum Communications, Inc. The license we filed for was held by WPIX, an independent station owned by what was then one of the nation's most powerful and successful newspapers, the New York *Daily News.* The *News,* in turn, was owned by Chicago's Tribune Company. Our quixotic but at the same time very professional and experienced group was determined to demonstrate how a responsible local television station should operate in a community that was being consumed by racial strife. Despite its elements of naive do-goodism, our license challenge offered a realistic and carefully prepared blueprint of how to operate a commercial television channel in the public interest and still make money.

I remember flying to the nation's capital on the old Eastern Airlines shuttle to file with the FCC our elaborate station license application filled with community surveys, business plans, and proposed program schedules. I saw the familiar and inspiring white granite and marble monuments from the air as we started our descent to the airport—the Washington, Lincoln, and Jefferson memorials, the White House— and I felt thrilled at the prospect of being given the opportunity to make our case in an open court before a federal examiner. I was not totally unrealistic about the enormous political pressures and influence that would be brought to bear to save the WPIX license, which was worth tens of millions of dollars. I had spent more than a decade at the networks during the period of the quiz scandals and in my own business running political campaigns. So I knew a good deal about how the system worked. All I sought was the chance to show the industry our blueprint that detailed how a responsible local broadcaster should operate.

However, the more I became enmeshed in the influence peddling and legal corruption of the nation's capital—and of the FCC in particular—the more disillusioned I grew. The *Chicago Tribune,* we were told, had gotten to Illinois Senator Everett Dirksen, the powerful Republican leader, whose minion in the FCC, the Broadcast Bureau chief, secretly gave WPIX an unprecedented and completely unauthorized early renewal. Fortunately, his surreptitious action also turned out to be illegal. When the scandal was exposed in *The New York Times,* the comparative license hearing we sought before the FCC had to be granted, but Forum's license challenge was put through tortuous delays and bureaucratic entanglements—anything to avoid a decision on the merits. The comparative hearing against WPIX compiled what was at the time the longest record in FCC history. Forum won the first round but the appeals went on for years. The case was still plodding along in 1976, when I left to become president of PBS. Several years later, the remaining members of the Forum group withdrew the challenge in return for a substantial financial settlement.

WPIX kept its license, promised to reform its ways, and replaced its chief executive officer with a dedicated broadcaster.

Reporters who cover cases like ours before the FCC and other regulatory agencies know all about the influence peddling, lobbying, and manipulating that goes on every day to protect special interests. Because their alienation tends to be born of inside knowledge, the reporters' distrust of government is no mere matter of generational conflict or innate professional cynicism. They are more deeply affected than most citizens because they know more about what goes on inside government than most citizens. But based on recent survey data and election results, the vast majority of the nation's citizens of every political stripe have come to share the journalists' conviction that their government is not deserving of their trust.

More influential than the personal and political leanings of individual newsmen and women is the growing impact of the "corporatization" of American journalism. The great trend in newsrooms today is toward increasing conformity. The mainstream media tend to reinforce and reconfirm mainstream values and establishment views, not in response to the orders of their media owners and corporate employers but because the news professionals themselves identify with those values and beliefs. Journalist Jack Newfield pointed out that the content of most news stories rests on an accepted set of political givens that are unquestioned and unchallenged: "The men and women who control the technological giants of the mass media are not neutral, unbiased computers. They have a mind-set. They have definite life styles and political values, which are concealed under a rhetoric of objectivity. But those values are organically institutionalized by *The Times,* by AP, by CBS . . . into their corporate bureaucracies. Among these unspoken, but organic, values are belief in welfare capitalism, God, the West, Puritanism, the Law, the family, property, the two-party system, and perhaps most crucially, in the notion that violence is only defensible when employed by the State. I can't think of any White House correspondent, or network television analyst, who doesn't share these values. And at the same time, who doesn't insist he is totally objective."

"Given the striking similarity in the private political and economic goals of all the owning corporations," journalism analyst Ben Bagdikian has warned, "it is not particularly comforting that the private control [of the major daily newspapers, magazines, television, books, and motion pictures that reach the majority of Americans] consists of two dozen large conglomerates instead of only one." These conglomerates are the media gatekeepers that wield extraordinary power over the ideas and information the public receives. While the total number of TV channels and media outlets is burgeoning, the number of owners in control of the nation's mainstream media, both print and electronic, is shrinking rapidly. Media mergers and acquisitions are producing an intensely concentrated information industry and a big business climate in which those who work for mainstream news organizations are made very much aware of the corporate limits within which they are expected to function. Iconoclasts, nonconformists, and unconventional thinkers at either end of the political spectrum do not thrive in today's multi-billion-dollar corporate media environment.

Direct corporate intrusion into news content is rarely necessary and in any event has largely gone out of fashion with the rise of the corporate mentality throughout the editorial side of the mainstream media business and the passing of the idiosyncratic media barons, like William Randolph Hearst, Colonel Robert R. McCormick, and Joseph Pulitzer.

There are exceptions, however, especially when the interest of the owner is involved. For example, during my first years as president of NBC News, I had virtually no direct involvement or contact with NBC's owners upstairs at RCA regarding the content of our news programs. Six decades earlier, NBC had been founded as a programming service to promote the sale of RCA radio sets. Over the years, RCA had seen NBC mushroom into its most profitable business by far. Extremely sensitive to possible antitrust action, the parent company made it a strict policy to keep an arm's length distance from its powerful broadcasting subsidiary. And that distance was doubled for news, which functioned as a free-wheeling, highly independent public service operation, largely detached from the company's business interests.

After GE bought RCA in 1986, however, the climate changed radically. Early on the morning of Tuesday, October 20, 1987, I received an angry, sputtering phone call from my new boss, GE chairman Welch, a chief executive noted for his intensity and quick temper. GE had recently bought NBC's parent company, RCA, and in a major break with tradition, Welch had quietly added the duties of NBC chairman to his own GE portfolio. In the deregulatory climate of the eighties, Welch—unlike his RCA predecessors—felt no inhibitions in dealing directly with GE's new acquisition. He called me that morning in high dudgeon to complain about how NBC News had reported the sudden stock market collapse of the day before. Welch insisted that our newsmen stop using phrases like "Black Monday," "alarming plunge," and "precipitous drop" to describe Wall Street's October 19 debacle. Their reports were only making the situation worse, Welch said, scaring stockholders and further depressing the price of even the healthiest companies, like GE, NBC's new parent. NBC News should stop fouling its own nest, he told me. Welch and I had a rather heated exchange about the appropriateness of his editorial interference, which had caught me by surprise. I told him I thought it would be prudent for both of us to keep our conversation between ourselves. I did not plan to pass along his comments to the NBC News reporters. It was the only such phone call about news content I ever received from him. By the following year, NBC and I had parted company.

Even more revealing of the corporate values of many of today's media owners was the memorandum dispatched the following month to NBC executives from its new president Robert C. Wright. Wright had been posted to NBC from GE, assigned by Welch to replace Grant Tinker, who resigned right after the corporate changeover. In the memo, one of Wright's first to NBC managers, he announced that GE had decided to form an NBC political action committee, to which the members of the network's management team were expected to contribute generously, since NBC's future depended so heavily on communications policies made in Washington. "Employees that earn their living and support their families from the profits of our business must recognize a need to invest some portion of their earnings to ensure that the company is well represented in Washington and that its important issues are clearly placed before Congress," Wright's memo said. "Employees who elect not to participate in a giving program of this type should question their own dedication to the company and their expectations." GE's managers have only one preoccupation—not the quality of the network's programs or the character of its journalism, but the bottom line. The news division responded by unilaterally declaring itself excluded from GE's new NBC corporate PAC and exempt from making any payments to it. As I wrote to Wright, it was entirely inappropriate for newsmen and -women to grind a political axe even for their own company. In the rather understated words of Yale law professor Stephen Carter, "We are moving into a world in which information is controlled increasingly by those who are not totally disinterested in the outcomes produced by the system."

In *Who Will Tell the People: The Betrayal of American Democracy*, journalist William Greider argued that the dissemination of political information today is dominated by large media conglomerates and lobbyists for special interests who can well afford to spend whatever it takes to promote their own interests and, even worse, to do so at the taxpayers' expense with tax deductible funds. Companies and special interests thoroughly outspend public interest groups and individual citizens who, by contrast with the corporate sector, have far fewer resources and far less incentive to spend their own time and money on politics. We shall go into this in more detail in the second half of the book. But as philosopher John Rawls wrote, "The liberties protected by the principle of participation lose much of their value whenever those who have greater private means are permitted to use their advantage to control the course of public debate. . . . When parties and electors are financed not by public funds but by private contributions," Rawls continued, "the political forum is so constrained by the wishes of the dominant interests that the basic measures needed to establish just constitutional rule are seldom seriously presented."

In contrast to a party press or ideologically driven media, what all the mainstream media have in common—whether crude newspaper and television tabloids (like the *New York Post, Hard Copy,* and *Geraldo*) or high-minded journalistic endeavors (like *The New York Times,* the *Washington Post,* and *This Week with David Brinkley*)— is that they are essentially marketplace-driven enterprises. To be fair to the latter group, however, it must be said that they are produced with an admirable dedication to the public interest and a good deal of respect for their special role and responsibility in a democratic society. The mainstream media, by definition, cater to specific audiences and therefore tend to share their readers' and viewers' existing attitudes and views. Rarely do they veer from the path of mainstream thinking to advocate unorthodox views or radical ideas either of the left or the right. They tend to respond to and reinforce public opinion, rather than shape it and lead it. Their effort, understandably, is to interest and please their large audiences rather than risk alienating or offending them.

For that reason, the mainstream media are a difficult environment in which to launch new ideas or gain acceptance for new and unknown faces. But once there is a sense that an idea or a personality is starting to take hold, to "make it," the media climb on the bandwagon and accelerate the new entry's visibility and popularity. Every major recent political and social change—the civil rights movement, the women's movement, the anti-Vietnam movement, the rise of the evangelical right, the phenomenon of Ross Perot—at first was largely unrecognized and ignored by the mainstream media. Only after the movement had reached a critical mass of acceptability was its existence recognized and promoted on television and in the press.

Mainstream news, as might be expected, caters to those institutions, like the government, that are in the best position to supply news. The structures of power that hand out press releases, stage press conferences, deliver statements, and stage events or—to use Boorstin's phrase, "pseudo-events"—get the lion's share of attention and coverage. Reformers, critics, those on the edge of society with unorthodox or unpopular ideas, get short shrift. When I was recruited from PBS by Tinker to leave Washington, D.C., and return to New York to run NBC News in 1984, I came without a specific professional background in television news, although I had long experience both at PBS and as an independent, producing and overseeing public affairs series and specials. Like most people, my image of a major news organization was of an aggressive, enterprising army of reporters, producers, and news crews who traveled the world searching out stories and discovering new developments to cover. I was in awe of the network's large, elaborate, and expensive worldwide newsgathering capacity, with bureaus and stringers in major cities on almost every continent. I spent my first two months at NBC News as an observer, learning the ropes, finding out how the news-gathering system worked from my far more experienced predecessor, Reuven Frank. To me, the single biggest surprise of that period was the fact that NBC News and, as was quickly apparent, all the other leading news-gathering organizations, spent almost all of their time and resources reacting and responding to news events scheduled and staged by others. We initiated virtually no truly original reporting of our own. The time of all of our daily hard newscasts was entirely filled by news conferences, speeches, celebrations, and breaking news scheduled or staged by governments, public officials, high-profile personalities, front-rank corporations, interest groups, and public organizations like the United Nations, the Red Cross, the political parties, and pro-choice or anti-abortion advocates. I could find hardly any news that we ever "dug up" ourselves except for an occasional soft feature story or oddball color piece, and even most of them originated from press releases, public information directors, or publicity agents. NBC News, like all other mainstream news gatherers, was being programmed almost entirely by others.

In an effort to recapture the news initiative at least some of the time, NBC News started to focus on week-long themes in which every network news program would produce a major story on different aspects of a particular issue—the cold war, education, the environment, the global economy, the aftermath of Vietnam. Our goal was to gain news leadership by providing more in-depth coverage and stress original reporting. It did not take long, however, before the press of daily deadlines, budget constraints, and the need to cover the same major unfolding stories as everyone else took us back to the practice of reacting to the themes and priorities of others—mostly what the president, the Congress, and top government departments and agencies decided to announce as the news of that day. I had more opportunity to initiate stories and commission programs with original themes at PBS, which had no hard-news-gathering capacity and produced no programs of its own, than at NBC News, which had thousands of newsmen and -women at its beck and call. I had never appreciated before just how much the demanding process of daily news-gathering helps to build an image of reality that only reinforces establishment attitudes and bolsters conventional viewpoints.

Another surprise I found at NBC News, which turned out to be characteristic of most other news organizations I came into contact with as well, was the striking absence of intellectual curiosity by the members of the news division concerning the policies and issues we were reporting on. The prevailing attitude among most newsmen and -women was almost *anti*-intellectual, disdainful of theoretical discussions and analysis of issues. The great majority of reporters, writers, correspondents, producers, and editors I worked with were doers, rather than students or thinkers, preoccupied with logistics, strategy, tactics, and process, rather than ideas, theories, or issues.

At academic conferences and seminars, journalists are fond of referring to their calling as a profession. But journalism is actually more of a trade or craft than a profession, which, as Webster's Third New International Dictionary defines it, requires "long and intensive preparation including instruction in . . . scientific, historical, or scholarly principles . . . committing its members to continuing study."

Reporters are always in a hurry and working on deadline. There are few scholars or intellectuals among their ranks. Examining the substance of what they are reporting, with the subtlety, nuance, and the intellectual background required to capture the subject accurately and effectively, is not the stuff of headline news. Brevity, clarity, and speed are the prized values. The trick is to look for the most exciting and most controversial elements, the areas of dramatic conflict: Who's

up, who's down? Who's ahead, who's behind? These are the easiest
guidelines for shaping a news story, and, above all, the fastest way to
write an interesting and coherent story.

After-hours discussions and bull sessions among journalists almost
never focused on the substance, ideas, or theories behind the issues in
the news, but on winners and losers, political and personal motivations, tactics and strategy, and, above all, personal gossip.

Many analysts have made the point that John Rawls articulated,
that "the tradition of objectivity in journalism has favored official
views, making journalists mere stenographers for the official transcript
of social reality." The issue is not intentional bias or conscious (even
unconscious) political slant, but the way in which the business of news
is carried out. Most often, it takes "outsiders" to find new priorities
and develop original news themes: Rachel Carson focusing on the environment in *Silent Spring* and Ralph Nader examining automobile
safety in *Unsafe at Any Speed.* Former *Washington Post* editor Ben
Bradlee credited his newspaper's leadership in reporting the Watergate
scandal to the fact that the story of the local break-in that started it
all had been assigned to two young reporters on the metropolitan desk,
Bob Woodward and Carl Bernstein, who were outside the journalistic
elite power structure and completely separate from the newspaper's
regular national staff.

An examination of the mainstream media's initial response to the
Ross Perot phenomenon of 1992 offers a good illustration of the
press's usual bias toward the predictable and against the unorthodox.
Hardly anyone in the press noticed or reported Perot's first election-
year appearance on *Larry King Live!,* Tom Rosenstiel reported in
Strange Bedfellows. "The *Los Angeles Times* ran a tiny item on page
18 two days after the King program about Perot's offer to run if people
in 50 states put him on the ballot. Two weeks after that, the United
Press International ran a small item about Perot's phone banks out of
Dallas. So did *The New York Times.* The *Washington Post* ran its first
small item at the end of March, a month after the King show. . . . By
the time the *Washington Post* and *New York Times* took serious notice of Ross Perot in late March, he was at 20 percent in the polls,
and he had done it without benefit of party, elections, or even attention
of the establishment press."

The mainstream media, led by television, are unremittingly conventional and cautious in their approach to new political ideas and public
policies. Their sense of themselves as "the fourth branch of government" suggests that they define themselves as part of the ruling estab-

lishment rather than as an independent force not constrained by the conventional wisdom. Whether liberal or conservative, up-market or down-market, the establishment media tend to accelerate and popularize existing trends rather than inaugurate new ones. Like the effect of wind on the tide, the media increases the intensity and velocity of whatever already has begun to gather force. They are pushers more than pullers of public opinion. As someone observed about television's vaunted influence and power, in reality it is a "sheep in wolf's clothing."

NEWS AS AN EPISODIC VIEW
OF REALITY

Using fresh headlines, pithy sound bites, and quick cuts, the restless spotlight of the mass media keeps moving to focus on the new, always on the new. The flow of the news emphasizes emergencies—disorder and crisis, rather than coherence and continuity.

What is news depends on what makes a good story. The news media's view of the world is influenced, and at times it might be said even distorted, by what Syracuse University Professor Thomas E. Patterson calls the "situational bias" of journalism. It is the "situational bias" of the news, more than any liberal or conservative partisan slant, that influences what stories get covered and how they are reported. Journalism's biases lie mostly in five areas: in favor of what is new, what is bad, what is dramatic, what is most readily available, and what can be readily understood.

By definition, while operating within its relatively orthodox framework, news focuses on what is different and unexpected rather than on what is familiar and predictable. What is new keeps changing all the time, so the news offers a peculiarly fleeting, episodic, and disjointed view of the world. "The press is like the beam of a searchlight that moves restlessly about, bringing one episode, then another out of darkness into vision," Lippmann wrote. "Men cannot do the work of the world by this light alone. They cannot govern by episodes, incidents and eruptions."

News is also usually about something bad happening—war, crime, corruption, violence, natural disaster. Good news is rarely reported unless—like peace breaking out, a welfare family winning the lottery, a victim's miraculous recovery, or a hurricane heading toward shore that suddenly veers off and blows out to sea—it represents dramatic change, includes a major element of the bizarre, or offers heartwarming human interest. As the cliche goes, "No news is good news," unless, of course, good news is bad news for somebody else. In the words of McLuhan, "Real news is bad news, bad news about somebody or bad news for somebody." Prosperity, tranquility, honesty, morality, decency, normality and on-time airline arrivals do not make news. Exceptions exist, of course. And there is even something of a new trend to look for positive news to report. On *ABC World News Tonight,* Peter Jennings has a regular segment, "American Agenda," dedicated to reporting solutions rather than only problems. And *USA Today* makes a promotional virtue out of its dedication to celebrating success. "Journalism of hope," the newspaper's founder Al Neuharth called it.

The very essence of journalism, however, is uncovering whatever is bad rather than reporting whatever is good—questioning motives, searching out immorality, focusing on controversy, and in general looking for the worst in people and situations. A content analysis of press coverage of recent presidential campaigns concluded that "the press debunks most candidates' promises, implying they don't intend to fill them or they couldn't if they tried." Despite the fact that studies show most campaign promises and platform pledges actually have been kept, "the press treats almost every campaign promise as a calculated deception." So people go to the polls convinced their only choice is the lesser of two evils.

To capture the audience's attention, news also focuses on and emphasizes the dramatic rather than the dull. It tells a story, after all, and feels compelled to make its stories interesting. Coverage tends to be anecdotal, centering on a few individuals or a specific incident, selecting images that make an eye-catching headline, an appealing lead, and a lively, simple story line. By singling out the most newsworthy, the most provocative, tendentious, and dramatic elements of any event, the press often exaggerates what it reports. Rarely is any news story ever underplayed. So reality is portrayed as a series of crises and cliff-hanger plotlines, as opposed to the plodding process of daily life. The job of the reporter and producer is to get into the paper, get on the air, and grab the public's attention. They do what they can to "sell" their story to their editors and to their audience by making it seem as newsworthy and important as they reasonably can, emphasizing the most controversial and most dramatic elements.

When an earthquake struck outside of San Francisco at 5:04 P.M. Pacific Standard Time, on October 17, 1989, television cameras inside the city, where the news bureaus are located, searched for the nearest scenes of the worst damage and kept showing them over and over again, thereby giving viewers the impression at first that much more of San Francisco had been destroyed than actually was the case. Proximity played an important role in the coverage. Even more severe damage in Oakland, for example, got relatively little early attention because virtually all the television cameras were in San Francisco, on the other side of the Bay Bridge.

When a riot erupts in Manhattan, Seoul, Johannesburg, or Jerusalem—even if it has been confined to a single street or neighborhood—the impression is left by news reports that the entire city is unsafe or in turmoil. My wife and I were in Seoul in the spring of 1988 when students staged marches in various parts of the city to protest the death of one of their dissident schoolmates at the hands of the police. The NBC News crews raced out of the bureau in full riot gear, with gas masks and crash helmets, to cover the violence in the streets. They came back covered with dust and gunpowder from the security forces' tear gas shells, but their reward was plenty of dramatic news footage powerful enough to make the nightly newscast. My wife went shopping in other parts of the city, which were completely calm, and returned with gifts for our grandchildren. The scenes NBC News showed on the *Today* show and *Nightly News* were filled with students and police clashing violently, giving the impression that all of Seoul was in turmoil.

I had a similar experience in Jerusalem in 1986, the week that the intifada—the uprising started by Palestinian rock-throwing youths—erupted. I was there to speak at an international symposium on freedom of the press, sponsored by Hebrew University. When the riots broke out, I divided my time between attending the symposium discussing free press issues with Israeli and American judges, lawyers, and scholars, and traveling with the NBC News crews covering the demonstrations. The outside world saw the riots, tire burning, head splitting, and rock throwing that seemed to engulf all of Israel. The reality was much more of a mixed picture; most Israelis—and Palestinians for that matter—went about their daily lives unaffected by the nearby pockets of violence and unrest.

During my visits to hot spots, like Seoul, Jerusalem, and Johannesburg, people I met at cocktail parties and dinners would invariably ask how my wife and I could live in New York City with all its crime and violence, which they saw night after night on their own country's television network. Would it be safe for their children to visit there? Or would it be better for them to stay home in the safety of Seoul, Jerusalem, or Johannesburg?

"The highest power of TV journalism," observed former NBC News President Frank, "is not in the transmission of information but in the transmission of experience—joy, sorrow, shock, fear. These are the stuff of news."

The "situational bias" of the news does little to provide a coherent framework for thoughtful public dialogue or to stimulate careful examination of sound policy alternatives. On television especially, the use of quick-cut images and short sound bites tends to foreclose reasoned reflection and time-consuming deliberation. Television's attention span is notoriously short, and in the MTV age, growing shorter. News is designed to capture instant attention in an intensely competitive and distracting environment, in which viewers with zappers "graze" through channels the moment their interest lags. The impulse is to excite and titillate, even exaggerate reality if necessary, to make stories as compelling as they can be. Television has no time for nuance or subtlety. The emphasis on metaphor and image heightens the impulse for immediate, intuitive response. As one journalist complained, "Television can give us so much, except the time to think."

The popular news media, led by television, perform an important service by calling the public's attention to certain issues and events. But they are seriously lacking in any effort to help audiences work through thoughtful judgments about policy alternatives. Complex issues like health care reform and educational policy that require sophisticated, time-consuming explanation and background get short shrift from most commercial mass media outlets. The episodic, negative, provocative, and simplistic proclivities of most popular news media end up narrowing rather than broadening the range of the public's choices.

The next chapter pursues in more detail the multi-faceted, fast-changing role of television. In its many guises, television, by definition, holds the key to the future of the electronic republic.

5

TELEVISION
AND BEYOND

No doubt the fact that it takes almost no effort to watch television is one reason it is so enormously popular. Of all the activities that people engage in, only "resting" has been found to be less demanding than watching television. Yet television is the most influential communications medium in history, its images the most widely received, its impact on our political system the most profound. While Americans spend half of all their free time with television, "the mean levels of concentration, challenge and skill were significantly lower for watching TV than the average levels for all other activities combined."

In any discussion of television and politics, four of television's special characteristics deserve to be singled out, if only because of their pivotal influence in defining, and in recent years, transforming the relationship between the people and the government:

1. Television gives ordinary citizens an unmediated, direct personal view of world events.
2. Television's presentation of reality happens in an environment dominated by entertainment and diversion, an environment calculated to help viewers escape from reality, not plunge into it.
3. Television emphasizes the personal; people and personalities take center stage over ideas and issues.

4. In a political system that is based on geography, television all but eliminates geography and replaces it with communities of interest.

THE DIRECT EXPERIENCE

Back in the 1950s, the affable chairman of the Republican National Committee, Leonard Hall, tried to explain why the traditionally rock-ribbed Republican voters of Maine suddenly started to vote Democratic, an inexplicable breach of faith to Republicans elsewhere: "Maine was always a Republican state," Hall said. "People were born Republican so they went to the polls and voted Republican. Except for Vermont, it was the only state to vote for Alf Landon over FDR in 1936. Then, suddenly, the folks up there in Maine started voting for some Democrats—too many from my viewpoint. So one day I asked an old Maine man what happened up in his state. 'Well,' he said, 'we can't do anything with this *television*. Our children were brought up to think that Democrats had horns. Now they see things for themselves on television, and realize that some of them don't have horns a-tall.'"

For the first time ever, people have an unmediated, firsthand view of the world outside of their neighborhood. On television "they can see things for themselves." In 1920, Walter Lippmann was certain that the press could not possibly supply citizens with the information they need to "acquire a competent judgment about all public affairs." Industrial society was expanding too fast and growing too complex. "The world we have to deal with," Lippmann wrote, "politically is out of reach, out of sight, out of mind. It has to be explored, reported and imagined." The basic problem, as he saw it, was that, like the parable of Plato's cave, the pictures inside people's heads do not automatically correspond with the reality of the world outside. To overcome "the limited nature of news . . . [and] the illimitable complexity of society," Lippmann proposed an elite "central clearing house of intelligence" that would inform both the press and the leaders of the government, and through them, the public itself.

Today, instead of Lippmann's elite guardians to serve as the central clearinghouse of intelligence, we have the electronic superhighways that lead directly into every government office and private home and provide instant and universal access to national political leaders and world events. To be sure, what travels on the electronic highways is limited, often flawed, and occasionally deceptive. But it is instantaneous, direct, and universal, capable of reaching everyone at the same time. Television offers intimate and often appealing views of much that was once distant and remote. It brings home major national and world events that formerly were available on a first-hand basis only to the privileged few.

In *No Sense of Place*, Professor Joshua Meyrowitz summarized that change. "At one time, physical presence was a prerequisite for first-hand experience. To see and hear a President speak in his office, for example, you had to be with him in his office. . . . What you read and heard was at best second-hand information. Live and mediated communications were once vastly dissimilar. This is no longer the case. . . . One can now be an audience to a social performance without being physically present; one can communicate 'directly' with others without meeting in the same place." Television's universal accessibility has produced an unprecedented sharing of information and experiences among all segments of the population, rich and poor, old and young, city and country, all of whom, regardless of class or educational level, have for decades been watching essentially the same programs.

What gives television its extraordinary impact in covering real world events is its ability to provide viewers with what seems to be an unfiltered, first-hand view, a direct experience. "The peculiar potency of television lies not in the wickedness of the journalists who operate the machine but in the very nature of the machine," a British official once remarked. It is a machine that gives the entire nation, and now billions of people throughout the world, the shared experience of participating in the major happenings of our time.

By contrast, the print media—newspapers, magazines, and books—provide essentially a second-hand view of the world, processed through the words of reporters, editors, columnists, authors, and public officials. Print journalists may be—and usually are—better informed, more experienced, and smarter about what they report than the average television viewer, or for that matter, the average television reporter. But the print reporter is still only a surrogate, and that's not nearly as good as being there yourself. Because television comes closest to putting the viewer at the scene, the public sees television as inherently more believable, more reliable, and more trustworthy than any other medium of information. That is demonstrated by the findings of every Gallup and Roper poll taken in recent years.

The special nature of television's appeal lies in its *coverage*, its ability to converge on an event such as a war, a political convention, an earthquake, or a presidential speech, and transmit what is happening while it is happening. The strength of newspapers and magazines lies in their *journalism,* their ability to transmit coherent descriptions that reconstruct, interpret, and describe what has already happened. To the public, live pictures that people can see for themselves inevitably are more real and reliable than someone else's secondhand written report, no matter how evocative or intelligent the writing.

That explains why television's most visible and most popular news stars, the national and local anchors, have so little influence on the public's views on any major issue. Tom Brokaw was not chosen as NBC News's ubiquitous anchor for the bright ideas, exceptional perspective, or depth of knowledge he might bring to the news. An ugly genius would not make a good anchorman, although he or she might make a great journalist. In fact, the vast majority of television's newsmen and -women serve essentially as little more than video page-turners and scene setters for the major news events. Their role is as a familiar and comfortable presence, a stand-in for the audience at home, rather than as authority or even journalist. Their popularity depends far more on their personal manner, appearance, and style of delivery, than on their intelligence, knowledge, or even their capacity to express an intelligent thought and write a coherent sentence. They "parachute in" to join the war in the Gulf, the UN mission to Somalia, the scene of an earthquake, the civil war in Bosnia or Rwanda. Most likely they've never been there before, know none of the local leaders or authorities, and possess almost as little background information about the story as the viewer at home. Having done their brief stint, they "parachute out" again, moving on to the next breaking story wherever it may happen. They don't actually "report," they "cover" what is happening.

In the game of show-and-tell, the preeminent power of television now lies with the "show." Journalism, the "tell," is secondary, almost nonexistent. As ABC's Ted Koppel remarked, "The electronic tail is wagging the editorial dog." Because seeing is believing, the camera rather than the correspondent has become the most powerful disseminator of news. The camera lens, of course, has its own inherent limitations. It favors close-up pictures and whatever can be seen in the immediate foreground. Cameras offer no clear view of what is going on in the background. They lack the capacity to depict depth and perspective and provide no sense of history. The camera requires viewers to make their own interpretations and judgments, framed by their own experiences and reactions, while watching the pictures on the screen at home.

During the early days of television, its network news correspondents were household names, famous for their articulate, thoughtful, and informed descriptions of the news they were reporting. Edward R. Murrow, Howard K. Smith, Charles Collingwood, Walter Cronkite, and a younger David Brinkley played a far more central role in the news picture than their successors do today, in part because the initial television news technology was comparatively crude and awkward. Picture and sound had to be recorded separately, one on film, the other on audiotape, and then painstakingly synchronized. Processing film was slow, and no communications satellites were in place to feed instantaneous reports live or on tape. Film reports had to be flown back to New York or California to get on the air. Pictures of the news were hardly ever available until a day or days after the event.

So the correspondent told the story. The very slowness in delivering the news gave television correspondents more time for thoughtful, analytical reporting than the jumping-jack coverage of today. The correspondents' narration accompanying their film reports during the Korean and Vietnam wars did more to add perspective and understanding than today's high tech, action-driven coverage that benefits from the use of satellite transmission, ubiquitous small-form cameras, instantaneous videotape and seamless electronic editing.

My own first experience with television satellites was typical of how many important technological innovations in communications have come about—more by accident, inadvertence, and backing-in than by vision, foresight, or bold pioneering. Initially, for example, CBS's imperious chairman, William S. Paley, urged the network's affiliates to stick with radio and delay getting into the new, untried, and, at the time, questionable medium of television.

When I came to PBS in 1976, I was briefed by its chief engineer, Dan Wells, about the new network's ambitious, pioneering plan to build television's first satellite distribution system, at a cost of forty-three million dollars. The Ford Foundation was instrumental in developing the satellite scheme, which called for PBS to bypass the telephone company's ground lines and transmit its programs to public stations through space via satellite. The argument was that satellite distribution would be cheaper in the long run, and that PBS—the newest and by far the poorest television network—would be better served relying on its own distribution system than leasing coaxial cable facilities from AT&T, even at the phone company's public service rates.

I thought the idea was totally foolhardy. I was not willing to have the network's lifeline depend entirely on a signal sent thousands of miles into the atmosphere to bounce off a tiny dish revolving in synchronous orbit with the earth. I questioned whether the untested new technology would work, and if it did, what would we do if the distant satellite should suddenly fail? PBS, which was fragile enough at that time, would be put out of business altogether. Most of my colleagues at the commercial networks, to whom I turned for their expert advice, dismissed the idea as a nonprofit organization's technological pipe dream.

I also thought that pouring forty-three million dollars into a fancy new distribution system would be a dumb distortion of public television's priorities. Why should the fledgling noncommercial system that I had come to head spend its scarce money and credit building the world's most advanced electronic superhighway when it had so little in the way of programs to send out? If the system could raise the money, I argued, we ought to invest it in quality programs first and worry about how they would be distributed later. Good programs were our reason for being, not fancy hardware. I had come to PBS trumpeting that my first three priorities would be "programming, programming, and programming." Building a pioneering satellite distribution system would only upend our programming priorities.

The network's chairman, Ralph Rogers, an enormously successful, public-spirited businessman from Dallas, who had recruited me to take the PBS job, was dedicated to the satellite project, as was the network's vice chairman, Hartford Gunn. In addition to the satellite's promised economic efficiency, they argued, public broadcasting's ownership of its own interconnection system would help insulate PBS from government control, always a potential worry—especially after the Nixon administration's almost successful efforts to choke off public television's public-affairs programming. The PBS chairmen had a point, but I still was convinced that their priorities were woefully misplaced and launched a one-man campaign to derail or at least postpone the entire PBS satellite project. Let somebody else try it first, I urged.

Then a new U.S. president, Jimmy Carter, was elected to the White House, and just before he took the presidential oath, I had the bright idea of trying to persuade the Senate to open its confirmation hearings of the new cabinet nominees to live PBS television coverage. For the first time in history the nation would be able to see its new leadership, the proposed secretaries of state, treasury, defense, the attorney general, and others, as they were being examined by the senators. This PBS initiative, I was convinced, would send a major signal that public television would be an important player in the nation's public affairs. The Senate leadership quickly gave its consent and we ran many of the confirmation hearings live.

Some public television stations applauded the new initiative as something of a programming coup. But many of the smaller stations around the country, especially those licensed to school boards and educational institutions, were furious. The televised hearings had forced them to abrogate their contractual commitments to schedule multiple daytime plays of PBS's educational children's programs *Sesame Street*, *Electric Company*, and *Mr. Rogers' Neighborhood*. These stations either had neglected to videotape the programs so they could rebroadcast them in place of the PBS feed, or simply did not have the equipment to do so. They were ready to go to war with the new PBS president, or at least to impeach him.

The proposed satellite interconnection system that I was fighting so hard to stop suddenly looked terribly appealing. Unlike the AT&T cable, the satellite would enable PBS to transmit several different programs simultaneously to its stations. Once it was in place, PBS could feed *both* the live hearings *and* the regularly scheduled programs, allowing the individual stations to choose which programs they wanted to carry. I became an instant convert to the magic of satellites and their ability to place more options at the local level while giving the network more programming flexibility.

Far from being a farsighted pioneer, I was simply trying to find a way to keep peace in public television and preserve my new job. So it was that public television, and not the far richer commercial networks, made history by creating the first satellite broadcasting system. After public broadcasting built its satellite distribution system on time and on schedule, and proved that it would work, the commercial networks quickly followed suit. When I moved on to NBC News in 1984, NBC was only just making the switch to satellite distribution. My years of experience with communications satellites at PBS helped to give NBC News an important early competitive advantage. Having learned to be open to the sometimes unpredictable advantages of new technologies, we also moved quickly to be the first to computerize the news division throughout the world and complete its conversion from slow film to instantaneous videotape.

Back in 1967, CBS News's Walter Cronkite, the most trusted man in America, had to fly all the way back to New York from his brief tour of Vietnam after the disastrous Tet offensive before he could appear on CBS Television to rebut the official forecasts of a soon-to-come American victory and deliver his own forecast that it seemed "more certain than ever that the bloody experience of Vietnam is to end in a stalemate." Live television reporting from Vietnam was not yet possible. And instant videotape coverage, which unlike news film requires no processing and developing, was not yet available. Today, Cronkite's piece would be done from the scene and most likely would be transmitted live to the American people who would not be nearly as interested in what Cronkite had to say as in what he had to show —so they could make their own judgments rather than have to rely on his.

AMERICA'S FIRST LIVE
TELEVISION WAR

The 1991 war in the Persian Gulf almost three decades later was television's first live war. Watching the continuous flow of pictures live from the scenes of the conflict, the world thought it was experiencing the war itself firsthand. Actually it was seeing an illusion of news, with almost none of the reality of combat and, until the very last days, almost no independent reporting. Too often, the on-the-scenes cameras and live satellite transmissions actually served to mask reality while apparently documenting what was happening. Even with the great volume of live picture and sound coverage by hundreds of inexperienced, transplanted local anchors, not to speak of dozens of experienced network news crews, truth continued to be the first casualty of war, as it always has been. In *Catch 22*, Yossarian, Joseph Heller's antic World War II airman, discovered that even what you are seeing for yourself with your own eyes can be dangerously deceptive, if not dead wrong. The experience of the nation's television viewers during the Gulf War was not much different.

The technological virtuosity of television's instantaneous nonstop coverage of the war disguised major journalistic flaws and deficiencies. Too many of the reports the nation saw turned out to be either totally inaccurate, only marginally true, or at best, a partial view that distorted reality and skewed perspective. On the war's very first day, for example, CNN reported that all of Iraq's Scud launchers had been destroyed, that Saddam Hussein's elite Republican Guard had been "decimated," that allied bombers had wiped out the entire Iraqi air force, that the Iraqi Scud warheads were filled with nerve gas, that Iraq had nuclear missiles ready to launch, that Israel had decided to enter the war. Every one of those reports turned out to be false.

From the Gulf, CNN and the other networks' live coverage transmitted rumor. Reports were broadcast without verification, without identifying sources, without editing: "Was that the sound of thunder or an incoming lethal rocket attack?" an anchor-reporter on the scene in Tel Aviv asked, while the camera quivered and viewers watched and wondered about the loud noises they, too, could hear live in the background. "Was that nerve gas or conventional explosives from a Scud missile?" Smoke could be seen seeping in front of the television camera. It turned out to be only bus exhaust. One veteran correspondent later wrote, "The electronic tools of the trade somehow transfixed [CNN] . . . into thinking it is free from having to use its journalistic tools of the trade." The coverage had the appearance of authenticity when viewers at home saw reporters in the Gulf scrambling for their gas masks and blanching at missile alerts even as they were delivering the "news."

The impressive daily briefings delivered by military officers and transmitted live to the audience back home were authentic, but the truth never had time to catch up to the inaccuracies. In their impatience to get on the air live, rather than wait to verify and digest what was actually going on, reporters stood in front of their cameras and speculated on what they were seeing and experiencing. They were not so much reporters, doing the job of filtering out truth from falsehood for the audience back home, as sideline observers, almost as ignorant of what was actually happening as their viewers in the United States.

These were not rare mistakes made in the heat and confusion of battle. Nor were they inconsequential gossip and rumor being whispered by small-town idlers on local street corners. These were authentic-appearing, real-time sound and pictures, transmitting major news about life-and-death matters to the entire world via satellite. The Gulf War's live television reports enthralled millions, not only millions of ordinary people but also high-level political officials and military officers who were responsible for making critical policy decisions. Secretary of Defense Dick Cheney acknowledged on the eve of the Gulf War that, like most Americans, he was getting most of his information about what was happening in the Gulf from CNN, an endorsement CNN was quick to advertise. After the war, General H. Norman Schwarzkopf, the commander of the multinational allied forces in the Gulf, revealed that he had "basically turned the TV off in the headquarters very early on because the reporting was so inaccurate I did not want my people to get confused."

Only long after the Gulf War ended did the American people learn that the smart bombs they had seen on television operating with unerring precision comprised only 7 percent of all the bombs dropped on Iraq and Kuwait. Ninety-three percent of the bombs the allies deployed were old-fashioned, unguided explosives that missed their targets 75 percent of the time. Only after the war was won was it revealed that the vaunted Patriot missiles, which on television were seen to "kill" almost every Scud in sight, had actually failed to destroy a single Scud warhead. When the Patriots hit anything, it was usually the missile's detached, used-up fuel tank. Far from being surgically precise, the bombing had caused "nearly apocalyptic" devastation, blasting parts of Iraq back to the pre-industrial age, according to a report by United Nations observers. Israel never did go into that war, nor did Iraq in reality have any nuclear missile capability.

There is nothing new in the fact that news reports from faraway places are often wrong. What is new is the fact that they are wrong even when buttressed by the unassailable evidence of live, on-the-scene pictures. For people who can see for themselves a live briefing from the war zone, a presidential speech, a race riot, or a close-up scene of children starving in Ethiopia, the reporter's role is significantly diminished. If I can see it for myself, I am not nearly so dependent on what the experts and analysts are trying to tell me.

In 1993, explaining the jury's decision to convict two Los Angeles police officers of using excessive force while arresting Rodney King, the jurors in the second trial said it was what they saw on the videotape rather than what the experts testified in the courtroom that persuaded them. "What we decided is to chuck all [the experts'] opinions. We said we're going to interpret [the videotape of the beating] ourselves. That's our job. We're not going to be bamboozled."

The public's increasing insistence on seeing things *for* themselves and figuring out things *by* themselves suggests a fundamental change in the way the American political system now works. Today, "the prime controllers of long-term public opinion are the Americans I call bystanders," as opposed to "the politicians and other persuaders . . . , pundits . . . , columnists, commentators, experts, lobbyists, spin directors and flacks," Columbia University sociologist Herbert J. Gans wrote in 1989. The "normally politically uninvolved members of the general public" are the ones who, like the King jurors, are making it a practice to shape the national viewpoint on their own, through their own bottom-up perspective, derived largely from what they see on television.

That fact was brought home to me most dramatically by a small journalistic controversy at the 1984 Republican convention in Dallas, soon after I became president of NBC News. To inject drama into an otherwise cut-and-dried convention that was going to do the utterly predictable—nominate Ronald Reagan for a second term—the Republican high command spent half-a-million dollars to produce a colorful, patriotic, emotionally charged eighteen-minute film, "Morning in America." They planned to use the film as a dramatic and innovative introduction to the president immediately before his appearance on the convention platform to deliver his acceptance speech. It was calculated to whip up the excitement of the delegates on the convention floor and, even more important, to overwhelm the television audience watching at home. The "Morning in America" film would be a far more effective introduction to President Reagan, they figured, than just another old-fashioned, overwrought convention speech delivered by one of the party's political luminaries.

Skillfully narrated by an experienced actor, the president of the United States himself, "Morning in America" opened with Reagan's taking the oath of office for his first term, followed immediately by images of the sun rising on a farm, a rooster crowing, a tractor starting up—the dawn of a new day in America. Bedecked with flags and with images of Americana—farmers, workers, an elderly couple, blacks, beautiful children—the film recalled the assassination attempt on President Reagan and his gallant recovery. All this was played against a song called "God Bless the USA."

Should NBC News broadcast this blatant campaign propaganda as part of its live convention coverage? Democratic Party Chairman Charles Manatt, fearful of the highly persuasive emotional appeal of "Morning in America," tried to stop it. He wired the networks, with copies to newspapers and wire services, insisting that we should refuse to televise the Reagan film. Manatt reminded us that at the Democratic convention in San Francisco a month earlier, the networks had totally ignored the Democrats' much less ambitious film profile of their presidential nominee, Walter Mondale. It was not a very persuasive argument since the networks had chosen instead to broadcast Senator Ted Kennedy's effusive speech introducing Mondale—as Republican Party Chairman Frank Farenkopf was quick to point out to us in a telegram of his own. "Morning in America," Farenkopf said, was the Republicans' high-tech equivalent of the Kennedy speech. Manatt's plea, coming on a slow news day in the middle of summer, put the dispute with the networks on the front pages of the nation's newspapers. Nothing else big was happening in the world and newspapers love to highlight problems that the television networks face.

The NBC News all-star convention team of Tom Brokaw, John Chancellor, and Roger Mudd rarely agreed with each other on anything and even more rarely ever talked to each other offscreen. I attributed their personal animosity to each other to star rivalry, which is pretty much standard for the business, whether in entertainment or news. (At PBS, Bill Moyers, Robin MacNeil and Jim Lehrer, Louis Rukeyser, and Bill Buckley felt the same way about each other, and would appear together only under duress, if at all. I already had a good deal of experience with the peculiar temperaments of serious TV news prima donnas.) But all three NBC News stars agreed with Manatt that since the Republicans' film was propaganda, not news, it should *not* be broadcast on NBC. They proposed that while "Morning in America" was being shown to the delegates in the convention hall, they would offer a news report about the film and the controversy it provoked. To make the Republicans' political propaganda part of our news coverage, they argued, would be to make NBC News, in effect, an agent of the Republicans' political artifice.

The newsmen had a point, of course, but the film itself had by then become the day's biggest headline news thanks to Manatt's miscalculation in going public with his protest, and it seemed to me that the public would be eager to see for themselves what all the fuss was about. Why not let the NBC viewers see "Morning in America" and tell them how the film came to be made, I suggested, rather than try to protect our audience from it? I also pointed out that if the Republicans had refused to let NBC News televise their film, arguing that they had made it to be seen only by convention delegates and not the general public, our anchors would have charged them with censorship. NBC News would have moved heaven and earth to overturn the Republican-imposed ban, all the while proclaiming "the public's right to know." If the Republicans had really been smart, that is probably exactly what they would have done, while slipping us a copy of it under the table. That would have virtually guaranteed that the networks would run the film.

"Morning in America" rapidly escalated into such a big news story that I decided to override my more experienced news colleagues. I concluded that if I were an ordinary television viewer, instead of a newsman worried about protecting my image of independence, I certainly would want to see the film that was generating so much controversy. I would have been furious if the network had deprived me of the chance to view it for myself while Brokaw, Chancellor, or Mudd tried to describe what the convention delegates in the hall were seeing, instead of simply showing the film as part of the live convention coverage. "Morning in America" was the biggest news of the day; why try to protect people from it? I felt it would have been senseless to refuse to show it on television.

NBC News aired "Morning in America" from beginning to end, while it was being shown to the delegates in the hall, as part of the network's regular convention coverage. Our news team also prepared background reports about the film, analyzing its content, describing the political strategy behind its use, and explaining the reason for the controversy. On the appointed day, Brokaw set the scene: "This will be an evening of scripted, colorful pageantry," Brokaw said, "kind of like an old-fashioned MGM musical, in which thousands of people and bands and balloons and confetti will move right on cue directed by an unseen hand. And at the climactic moment, Ronald Reagan, just like his good old friend Fred Astaire, will glide into view." It was a bit of gratuitous editorializing, but at least we put the film on the air so that people could decide for themselves whether it resembled an old-fashioned MGM musical or was simply an effective new form of political image-making.

NBC News was the only network to televise "Morning in America" and the audience responded by tuning to us in record numbers. The viewers wanted to see the real thing, not what Brokaw or any other news correspondents had to say about the film's purpose and its manipulative qualities. Eager to see "Morning in America" with their own eyes, the people were able to judge it for themselves without our so-called help. The ability to provide a direct, unfiltered experience is the unique power of television, one of its greatest assets, and also one of its great dangers.

In 1963, the year that the network evening news went from fifteen minutes to a half hour, the Roper survey of public attitudes toward television found that for the first time, television was listed as the chief source of news by the majority of those surveyed. By 1974, 65 percent of those surveyed mentioned television, while only 47 percent mentioned newspapers. In that year, too, for the first time, the majority of college-educated people surveyed mentioned television rather than newspapers as their main source of information about the world. Today, television has a commanding lead among all segments of the population as the main source of news and information. And in many households it is virtually the only source.

Thanks to television, the pundits' words may influence what people think *about,* but no longer do they have much influence on what people think. The public finds out what to think about mostly through television, makes up its mind based largely on what it sees for itself on television, and then discovers what it is thinking by watching what the polls report on television. The result of this somewhat circular exercise is government by popular consensus, with the public's elected representatives carefully monitoring the opinion polls both before and after every action they take.

With so much of politics now taking place in continuous view of the public on CNN and C-SPAN as well as on broadcast television, political compromise has been made more difficult than it was in the days when government officials could work out decisions among themselves in the cloakrooms and closed committee rooms of the Capitol. Once a position is taken in public it tends to become entrenched and issues become polarized. "There isn't enough wiggle room" anymore, former House Minority Leader Bob Michel complained, referring to television's intrusion on the lawmaking process. And now that Congress has opened committee meetings to C-SPAN and put all documents electronically on-line, the "wiggle room" is even less. At two-thirty in the morning on January 5, 1995, as the House of Representatives finished its opening-day session of the new term, the new Majority Leader, Republican Newt Gingrich of Georgia, thanked the members of Congress for working so late and also thanked the nation's viewers for tuning in to watch the members at work. It was confirmation of sorts that the people have, indeed, become the fourth branch of government. As William Safire wrote, lobbyists have been stripped "of their most valuable asset: exclusive access to information."

One problem with the new openness and eagerness to include the public in the process: Efforts to settle disagreements by splitting differences and negotiating compromises behind the scenes—the very essence of politics—are viewed by segments of the public as selling out principles to which their leaders had made firm commitments. That attitude is frequently reinforced by reporters, often so intent on not appearing to be "used" and on showing off their insiders' insight, that the first thing they do is cast doubt on motivations and treat any compromise as a political retreat or defeat. One consequence, ironically, is that while television enhances both the visibility and the "bully pulpit" of the presidency by keeping the president on the screen front and center, it also has made it increasingly difficult for presidents to lead. The president's high public visibility narrows his policy options, limits his "wiggle room" to negotiate, and reduces the time he has to put his programs in place and bring the people on board.

It is no accident that during the era of television's dominance we have had "gridlock government," the first presidential resignation, and three one-term presidents so far. Only Ronald Reagan, a professional actor, has served two full terms. In contrast to earlier days, when presidents could lead based on their own strong conviction of what had to be done, the tendency in the era of television is for a president to feel compelled to keep in step with public opinion as the people follow the major events on television day in and day out.

THE ENTERTAINMENT ENVIRONMENT

Television's coverage of politics is frequently criticized for its unseemly emphasis on entertainment values that too often drive out meaningful discussion of the issues. In 1992, candidate Bill Clinton got more mileage out of joking with a raunchy morning radio talk show host and playing saxophone and wearing dark shades on a late night television talk show than discussing his ideas on health care or welfare reform. Of course, "bread and circuses" have been staples of politics since the time of ancient Rome. And although historians and political scientists focus on the serious issues that make or break governments and empires, entertainment and celebration have been central to every election campaign and political event since the American nation began. During the first hundred years of the republic, politics provided communities with their main source of public diversion: elections meant rallies, bonfires, barbecues, bandwagons, parades, colorful decorations, speeches, debates, and free-spirited celebrations of every kind.

For most of the last century, however, presidential candidates kept above the fray. They *stood* for election, rather than *ran* for election as they do now, in the belief that presidents should preserve the dignity of the high office by staying off the campaign trail. To take the high road in politics meant that presidential nominees should remain at home while letting their political supporters do their dirty work for them. Campaign workers and political party operatives felt under no constraints or inhibitions to whoop it up at campaign time, making sure that party politics were not just serious political business but also fun and games designed to engage and involve the entire community.

Now that traditionally loose and personal system of local politick-ing has all but disappeared. Instead, we depend on the media, specif-ically television, to shape the public's political impressions. "Television news powerfully influences which problems viewers regard as serious," Shanto Iyengar and Donald R. Kinder reported in their study of tele-vision and American opinion. "By priming certain aspects of national life, while ignoring others, television news sets the terms by which political judgments are rendered and political choices made."

In making the transition to the mass media, especially to television, presidential politics and national public affairs coverage at first adopted a relatively solemn and dignified demeanor befitting the im-portance of the office. Much of the popular carnival atmosphere of traditional democratic politics was eliminated. At the networks, a rigid line separated news from entertainment; news was considered serious and important business. In newspapers, politics could be found in the news and editorial sections, rather than in the entertainment, sports, and feature sections.

Although network television began by being serious about news, the public's electronic view of politics has consistently been seen in a television environment saturated with comedy, drama, fiction, vio-lence, sex, gossip, and commercialism designed to appeal to the emo-tions more than to reason. Presidential messages urging financial sacrifice, or governmental reform, or national support for sending American troops into trouble spots abroad are sandwiched between entertainment shows and commercials for Frosted Flakes and blue jeans.

As the authors of *Watching America,* a comprehensive study of television entertainment, concluded, "The medium creates a kind of hyper-reality, a shared fantasy world that merges with and sometimes even replaces the more mundane world of real life. . . . Political con-troversies are played out in TV movies as quickly as news stories can be transformed into screenplays. . . . Docudramas blend reality with fantasy as quickly as the real world can provide dramatic material." The study concluded, "Today, its eagerness to tackle the very latest controversy has made television entertainment an integral part of the trend-making machine that entwines both news and entertainment media."

Today, the fire wall that separates politics from entertainment has all but disappeared. And even within news broadcasts, solid reporting about serious subjects is diminishing. The increasing entertainment orientation of television news, national as well as local, compels producers to seek out the most vivid settings—the artful "picture ops," featuring waving flags, glittering uniforms, or dramatic scenes of violence and destruction—even if the images being shown make the news producers direct accomplices in political manipulation and artifice. As pop artist Andy Warhol once observed, it is no longer easy to know "where the artificial stops and the real begins." Produced news reports have shifted from focusing on the words of candidates and political figures to concentrating on their images and actions.

The long-standing journalistic rules against staging, faking, and dramatizing reality are being fogged over in an intensely competitive media atmosphere whose standards are set by the worlds of business and commercial entertainment. Whatever succeeds in bringing in the biggest audience is not only acceptable but welcomed. In 1992, NBC News planted in a GM pickup truck hidden incendiary devices rigged to burst into flame when the truck was sideswiped on a test track. The rigging was done for a feature story exposing the danger of certain GM trucks whose gas tanks allegedly were badly designed. When GM called the network on its deception and threatened a lawsuit, NBC News explained that it was simply "enhancing reality" by videotaping the faked conflagration scene. It was a transparently deceptive journalistic practice that would have been utterly unthinkable just a few years earlier, before network news and entertainment had become so thoroughly integrated and intermixed that viewers had a hard time knowing where truth ended and fiction began.

Television gives its viewers direct access to reality while at the same time making them especially vulnerable to manipulation and illusion because its images seem so authentic; "pictures don't lie." With the blurring, some would say the erasing, of the boundary that separates what is genuine from what is contrived, politics on television has taken on many of the characteristics of entertainment and many of the techniques of marketing and advertising. It has become a cliche to say that presidential candidates are being marketed like bars of soap and boxes of cereal. More important, one can say that soap and cereal are being marketed like presidential candidates, using different messages for different demographic groups, with emphasis on a product's personality, glamour, and sex appeal.

In an environment of mass entertainment, where facts and information take a backseat to images and emotion, where feeling replaces meaning and reason, it would be unrealistic to expect serious matters of state to take the high road of Jeffersonian reason and intellect. In politics, historian David Barber observed, the "elitism of expertise" has given way to the "equality of emotions." Everyone has feelings; only some have information. In reaching out to the mass of voters through the mass media, the temptation is to focus on style and personality rather than competence, "the show horse" rather than "the work horse." Politics is depicted as a sports event, a horse race, rather than the more realistic "crabwise movement from one compromise to another."

THE RISE OF
PERSONAL POLITICS

When Queen Elizabeth II visited the United States in 1976, soon after I came to Washington to run PBS, we were able to arrange to have PBS televise her formal and very elegant reception and state dinner at the White House. It was the first state dinner ever opened to the public through live television, quite a coup for the fledgling public television system. Alas, at the last minute a drenching rainstorm forced the White House to change its carefully orchestrated dinner arrangements, and PBS's big, bulky cameras were caught hopelessly out of position.

The unprecedented live television coverage of the state dinner for the queen was an unmitigated production disaster. The only glimpse that PBS viewers could catch of the queen's grand entrance into the White House party was a brief sight of the hemline of her long evening gown, below which could be spotted two ankles and a pair of sensible shoes descending the elegant front staircase. "The Queen looks positively ravishing tonight," exclaimed the retired BBC commentator, an expert on royalty's public events, who had been imported to describe the British side of the occasion. The PBS audience had to take his word for it; for the first several minutes nobody could see any part of the queen but her sensible shoes.

The next day, the press criticism was withering, deservedly so. As the new PBS president, I had a lot to answer for. But my mother, who along with millions of other Americans watched the great event on television at home was thrilled nevertheless. "It was like peeking through a hole in the sidewalk fence at a grand party to spot all the stars and great names," she said to me later. "All my friends were excited that you let us look in on the president and the queen from the outside and we felt lucky to see anything at all."

In 1986 Pope John Paul II came to the White House for a historic visit with President Reagan, which NBC News and the other networks covered live. Sitting next to the president during a photo session to memorialize the occasion, the pope extended his greetings to the American people and expressed his gratitude to President Reagan for his warm hospitality. As television broadcast the scene to the people of the nation and the world, its cameras caught sight of the President dozing off in his armchair while the pope was talking.

Television brings us up close to presidents, popes, and royalty and takes us to important places and great occasions where ordinary people have never been welcome before. It enables the public at large to see their leaders day after day, sometimes at great length, when they are off guard as well as when they are "on," when they are alert or dozing, charming or irritable, animated or bored. Seen up close, the most commanding and powerful figures inevitably lose some of their mystery. They seem more like regular people than one would expect. They have lost the "magic" that British political and social commentator Walter Bagehot said daylight must not be allowed to penetrate. In the United States, the "magic" vanished in the 1950s with President Eisenhower's heart attack when Ike's cardiologist, Paul Dudley White, gave the nation a detailed televised account of the president's bodily functions. We learn that presidents and popes suffer from hemorrhoids, allergies, and urinary disorders; that a president's wife has been treated for alcohol abuse; and that a president's children do not have the best relations with their father and mother. Politics, which has always emphasized the public persona and charisma of the leader, has become personal to a degree never before possible except on an intimate village level. The "magic" has indeed vanished.

"The most remote and illustrious men are met *as if* they were in the circle of one's peers," one television study concluded. Viewers come to feel they know the people they meet on television almost as well as they know their friends and associates. It is an environment that tends to undermine the mystique of the traditional "great leader." The image of great leaders—Charles de Gaulle, FDR, Winston Churchill—is of people who are larger than life, who have about them an aura of mystery. They appear now to be products of a bygone age. In contrast, those who have emerged more recently seem to be natural, informal, comparatively unprepossessing types who wear well in today's atmosphere of intimacy and continuing television exposure. We see them in jogging shorts and bathing suits. We know the style of underwear they wear. It's hard to deliver a commanding, passionate, eloquent, or rabble-rousing speech and expect to get a rise out of an audience of one or two people at a time, sitting or lounging in easy chairs, or even in bed, just a few feet away from the television screen.

When watching a speech on television, the tendency is to think at least as much about the speaker as about what he or she is saying. Words take second place to nonverbal cues, personal mannerisms, gestures, expressions, and overall appearance. The focus is on the communicator as much as on the communication. Television thrusts the personal into the center of the public realm—Nixon's sweating vs. Kennedy's smiling. The printed page, by contrast, "is a shield behind which human idiosyncrasy and frailty may hide." The reader cannot tell much about an author's personality or character simply by reading what he or she writes. But after merely an instant's face-to-face encounter on television, the viewer invariably forms a strong first impression about what sort of human being the person on the screen is.

In September 1987, as I sat in the NBC Newsroom watching Robert Bork, an obviously intelligent and experienced jurist, answer questions at the Senate Judiciary Committee's confirmation hearings for his nomination to the Supreme Court, I commented to my colleagues that I thought Judge Bork's haughty and disdainful bearing was going to give him more trouble than his political ideology. In an effort to help Bork's cause, his ardent supporter, Senator Alan Simpson of Wyoming, lobbed him a "softball," asking the nominee why he wanted to sit on the Supreme Court. Judge Bork responded that it was the intellectual challenge that appealed to him. His answer sounded arrogant and self-absorbed, insufficiently sensitive to the awesome responsibility to the public of a lifetime appointment to the U.S. Supreme Court. Bork's own uncompromising personality, revealed on television, had done him in. The nation's viewers were not going to flock to his support.

Since television has come on the scene, political scientists and other critics have chronically complained that politics focuses too little on issues and substance and too much on personality and style. Viewers pay more attention to the appearance and demeanor of candidates than to their political programs and beliefs. That is not an ignorant or irrational response. People know that words can lie or deceive but physical expression, personal bearing, character, and personality, they believe, are harder to fake. The perceived difficulty of disguising or dissembling one's personality—of acting in front of the television camera like someone you aren't—helps explain why people trust the impressions they get from watching television more than the information they receive from reading newspapers and newsmagazines. In that regard, the electronic media are more revealing than print. People have a good deal more experience making personal judgments by looking someone "in the eye" and taking their measure, even if he or she is thousands of miles away on television, than by analyzing the content and logical consistency of someone's published campaign platform or issue papers. After all, the written words were probably written by ghostwriters. Viewers don't need to be educated or knowledgeable in order to come to a personal judgment about which candidate they like better as a human being, compared to the requirements for making a sound judgment about which candidate has a better position on some important issue.

Election campaigns on television, therefore, are less about issues than about personality and style—and that's not all bad. A political personality does not necessarily have to be eloquent or even verbally articulate to come across well on the television screen. For the ordinary viewer, logical argument gives way to his or her gut reactions and personal experience in responding to people. When Jimmy Carter's presidency was in serious trouble, the president decided to deliver what his aides billed as a crucial "fireside chat," a talk from the heart about the nation's energy crisis and President Carter's own personal vision of the future. To emphasize the intimacy and sincerity of his presentation, the president shed the traditional suit and tie and donned a cardigan sweater for the occasion. Afterwards, people remembered nothing of what the president said, but retained a vivid image of what he wore. As one veteran political observer later wrote, "The fact that he wore a sweater while delivering the statement, . . . has become legend and, significantly, everyone to whom I have talked ascribed a complex meaning to the act."

In day-to-day life, first impressions are generally made on the basis of what might be called symbolic signposts—style of dress, general appearance, class of job, type of car, religious affiliation, and political beliefs. Then, the better we get to know someone, the more these symbolic signposts recede in importance. I might decide that I like you and trust you as a person and choose you for a friend, even if I disagree with your politics or don't like your taste in clothes. If I like your personality better and trust you as a human being more than someone else whose political views and political party I happen to share, I might decide to vote for you even though I disagree with your politics. People liked Ronald Reagan personally and gave him large majorities, even while the polls indicated that many of them did not approve of his policies.

The television camera's unblinking eye can strip away the public mask and reveal the private behavior and personal human reactions of public figures—the way they handle themselves—that at one time were known only to intimate friends, close colleagues, and immediate family. Personal mistakes, verbal gaffes, physical handicaps, hesitations, and occasional episodes of incoherence get revealed that, before television, were never seen in full public view. Like Franklin Roosevelt's braces and canes, either they were hidden altogether or they were cleaned up in the official retelling. Today, public officials can't control or deny what shows up live or on tape; it is harder to hide their misplays. President Bush's discomfort during the second televised debate in the 1992 campaign became obvious to millions of viewers when the camera caught him sneaking glances at his wristwatch. He couldn't wait for the thing to be over, and every television viewer knew it. Leaders seem more vulnerable, less authoritative, and more human than they once were when the public at large could learn about them only from a distance.

The stripping away of the curtain to reveal private lives has its dark side in the current preoccupation with sex and scandal and intimate details about public officials. In 1988, an extramarital affair that was spied upon and then relentlessly seized upon by the media destroyed the political ambitions of an early front-running Democratic presidential candidate, Gary Hart of Colorado. In 1992, the Clinton campaign had barely started when it was almost obliterated by a single issue, the Arkansas governor's alleged sexual liaison with a former nightclub entertainer. Ironically, in the interest of appearing to be evenhanded, the press moved from that scandal to embarrassing speculation about the private life of President Bush as well. Concentration on private lives has a way of lowering the level of public discourse; to nobody's benefit but the tabloids'.

The question once asked by party brokers and political insiders in the privacy of the backroom—"Is there anything in your past or in your private life that we should know?"—is now asked in public, with the answers seen by millions on television. Personal secrets have turned into public issues and, with increasing frequency, public scandals.

In politics today, what Tocqueville a century and a half ago called the "eloquent and lofty manner of discussing the great issues of state" has been displaced by "an open and coarse appeal to the passions . . . , abandon[ing] principles to assail the character of individuals, to track them into private life and disclose all their weaknesses and vices."

THE DISAPPEARANCE
OF GEOGRAPHY

The epigram "All politics is local" spoken by Tip O'Neill, longtime Democratic Speaker of the House of Representatives, is a hoary cliché of politics. Congressmen are elected on the basis of local election districts; our political system always has been organized geographically into wards, districts, counties, and states. But that is not necessarily true of television. For satellite communications, geographical boundary lines are irrelevant. Most of what is truly local about local television news is limited to weather and sports. With few exceptions, local news is "mostly an extension of the entertainment programs that surround it," concluded the *Los Angeles Times*'s Pulitzer Prize-winning television critic, Howard Rosenberg. "If I want nightly triple features of violence—endless coverage of grisly, blood-spattered offenses that feeds our paranoia about crime—I know where to find it. Local news."

Local television's lurid "body bag coverage" of crashes, fires, crimes, sex, and scandals—and its characteristically affable and inane news anchor "happy talk"—are largely interchangeable from one region of the country to another, from one market to the next. Media analyst Stephen Hess studied the content of 106 local newscasts from 57 television stations in 35 cities of all sizes, in all parts of the country. His overall impression was "its sameness." Local news is "news from nowhere," news without "a sense of place."

During the first three decades of television, news was dominated entirely by three national networks—ABC, CBS, and NBC—and aimed at a nationwide audience. Not until the 1970s did broadcasters begin to recognize the potential of local news as a profitable and comparatively inexpensive way to fill station airtime. When Forum Communications filed its license challenge to New York City's WPIX in 1969, the very first issue brought against it at the FCC hearing involved Forum's "financial unreality and fiscal irresponsibility" in proposing to produce an hour of local news a night on the station. Today, hour-long local news shows are the rule not the exception and rank among the stations' biggest revenue producers. But typically, television's local news has little to do with geography and even less to do with a sense of community. Shaped by national news consultants, individual stations' news programs have a homogenized, interchangeable character—the "look, sound and feel of Action News/Eyewitness News/Newscenter News"—and the formula is the same no matter where the program originates. "More notable than the absence of accent is the absence of place, a sense that this news is special to this locale," Hess found. Most local news directors are transients, moving from market to market like itinerant baseball players. Local news anchors speak without regional accents, are rarely familiar with local affairs, and typically have no personal or professional acquaintance with the nuances of local politics.

The absence of a real sense of localism in broadcast television and radio is matched and even exceeded by the newer medium of cable, whose dominant news, as well as weather and sports channels—CNN, C-SPAN, ESPN and the Weather Channel—are all national, with occasional local "break-ins" to accommodate local advertising. A bare handful of cable markets offer community news services, produced at the lowest possible cost. But in general, cable is an industry that serves local communities with national program services that are packaged in one central place and shipped in by satellite. Most public service cable channels, reserved for public, educational, and local government access, go unused for lack of funding. DBS (direct broadcast by satellite) services now being marketed have no local components whatever. The television signal goes from a central source to the satellite and back down to small receiving dishes outside people's windows, with no opportunity for local programming or community involvement.

Obviously, local newspapers, by definition, have more community identification than local cable systems and radio and television stations. But, according to James K. Batten, chairman of the Knight-Ridder newspaper chain, even that focus has lamentably declined. Local newspapers, he charged, have grown increasingly "disconnected" from their own communities. Urging his newspaper colleagues to reconnect themselves to their communities, Batten made the point that "Newsrooms often are overstocked with journalistic transients. . . . Their eyes are on the next and bigger town, the next rung up the ladder. . . . There is always the temptation to make their byline files a little more glittering at the expense of people and institutions they will never see again."

The most influential and largest circulation daily newspapers—*The New York Times, Wall Street Journal,* and *USA Today*—advertise themselves as national media, although *The Times* recently has spent considerable resources beefing up its metropolitan New York coverage. Even local weeklies, the traditional purveyors of neighborhood news, are being bought up by large national chains and filled with syndicated columns and features, making their "quirky individual-ism . . . harder to find."

In the media world, the audience's geographical identity, the tra-ditional basis on which the structure of government is ordered, holds little relevance any longer. Media markets do not coincide with polit-ical districts. The New York metropolitan television market, for ex-ample, actually spans parts of three states—New York, New Jersey, and Connecticut—and overlaps numerous towns, cities, counties, and political districts. Television programs are aimed at people according to their demographic characteristics rather than their place of resi-dence. When advertisers, politicians, and media experts refer to "com-munities of interest," they are not talking about place but about economic, social, and political characteristics and attitudes held in common. What consumers buy and how they vote have a lot more to do with age, sex, income, education, race, and ethnicity than where they live.

As suburbs blend into one another, as people move about and com-mute long distances to jobs, and as media markets displace political boundary lines, our geographical identity and sense of community rap-idly erode. One consequence of the fact that the nation's media struc-ture has grown out of synch with its political structure is that the federal system, as originally conceived in the Constitution, in a sense has been turned upside down. People today know more about and feel more involved with their president, senators, and congressmen than with elected officials in their state, city, and in many cases even neigh-borhood governments. What used to be distant and remote has be-come close and familiar; what used to be close and familiar has grown distant and remote. The media spend less time and effort covering local news from state legislatures, city councils, and individual neigh-borhoods and communities than covering the glamorous faraway hap-penings in Washington, D.C. News about the mayor of Newark, New Jersey, or a school crisis in New York City is of hardly any concern to the people of White Plains, New York, or Stamford, Connecticut, even though all are part of the same media market. What happens in the federal government in Washington, D.C., on the other hand, is relevant to them all.

The main consequence of the loss of geographic identity is the ex-pansion of a different kind of politics—special interest and single issue politics. The issues that shaped the federal Constitution—big states vs. small states, Southern states vs. New England states, farmers vs. city dwellers—no longer dominate the political debate. Instead, politics is increasingly centered on ethnic, religious, financial, age-related, and sex-related issues. Purely local and regional issues certainly have not disappeared. Far from it. Environmental, racial, economic, and myriad other local controversies are alive and well throughout the country. But today the most visible political factions coalesce around single interests—abortion, gun control, health care, welfare, crime—that supersede traditional political boundary lines. Communications define communities. And communities define politics. On that basis, communities of interest, rather than geographical proximity, are redefining the politics of our time.

THE COMING OF
THE INFORMATION SUPERHIGHWAY

What will define the politics of the future? Broadcast television, the mass medium that remains the most formidable and influential communications force, is being diluted by the rise of personal and interactive electronic media, the building of what Vice President Al Gore calls the "National Information Infrastructure," otherwise known as the information superhighway. One astute media observer described the new elements of that overused metaphor: "At one level, it's like having a highly efficient telephone and telegraph service with both private and party lines. At another, it's like flying an information spaceship and docking at other computers to copy lists and texts and pictures and other computer programs. And at yet another level, it can feel like an untethered walk in space to visit other travelers."

By combining elements of the telephone, home computer, television, cable, and satellites, one can send and receive oral, printed, or picture and sound messages; order television programs, pizzas, operas, and prize fights on demand; buy and pay for goods and services of all descriptions; send and receive bank statements, checks, and personal financial data; carry on all manner of financial transactions; send and receive information, reference material, articles, texts, books, catalogues, and CD-ROMs on any subject from almost any source, and hold private and multiple conversations individually or simultaneously with anyone anywhere in the world who can also plug into the electronic superhighway.

Some go so far as to suggest that the new personal telecommunications media soon will displace the present-day mass media, as cars replaced horse-drawn wagons and airplanes replaced long-distance trains. Predictions about the future of the media, however, while often hyperbolic, are always hazardous. Thomas Edison thought his new invention, the phonograph, would be used as a dictating machine. Alexander Graham Bell looked upon his telephone as an entertainment medium. *The New York Times* predicted that television was unlikely to catch on as radio did, because the size of the investment it would require "dwarfs anything to which even American capital is accustomed."

In Cerritos, California, a cable system that was hailed in 1989 as "the most sweeping test yet of interactive TV," a system that was supposed to "shape the future telecommunications systems for the whole country," was pronounced a bust four years later. "After a prolonged opportunity to serve as guinea pigs for the TV of tomorrow, hardly any residents subscribe," the *Los Angeles Times* reported. "I don't know of anyone who uses it," the mayor of Cerritos said. "For the average person in Cerritos, it doesn't exist."

Still, there is no doubt that we are in the midst of a telecommunications revolution whose characteristics are yet to be clearly defined. There will be no shortage of channels into and out of the home. In fact, in an age of video dial tone and digital transmission, there may no longer be channels—only unlimited bits of information and data to be translated into any format one wants to call up. With digital transmission, signal compression, fiber optic transmissions, and expanded use of the electronic spectrum, information of all kinds in all forms will pour in and out, limited only by people's ability to pay. It will travel via cable lines and telephone lines, through the air, and directly to and from satellites.

This interactive telecommunications revolution is already in the process of permanently and profoundly changing our nation's political system. "As the speed of information increases, McLuhan wrote, "the tendency is for politics to move away from representation and delegation of constituents toward immediate involvement of the entire community in the central acts of decision."

Trying to understand just how and in what ways that will happen will be the focus of the remaining chapters.

6

DECLINE OF
THE OLD POLITICS,
THE RISE OF THE NEW

As the general public grows increasingly involved in day-to-day decision making alongside the president and Congress, the role of the political parties, the chief instruments of the old politics, continues to decline. The major parties no longer dominate the nation's political processes as they have throughout most of American history. Besides raising money for candidates, the parties count for little today. Their national nominating conventions no longer nominate presidential candidates. Instead, the voters do the nominating themselves at the increasingly important state primaries and caucuses. The parties no longer carry the candidates' message to the people. Instead, CNN, C-SPAN, live television, talk radio, and political advertising put viewers and listeners directly into the political mix. The parties are no longer the chief point of contact between the electorate and the politicians. Instead, political interest groups reach out directly, using computerized mailing lists and modems in addition to the old-fashioned campaign techniques.

The parties have failed to adapt and have lost much of their power and influence and most of their reason for being. Party registration is down. Loyalty has eroded. Less than one hundred thousand members regularly contribute to the Democratic National Committee and the average contributor is more than seventy years old. The Republican national base is larger, but membership is still less than one million.

Local party machines have grown enfeebled. The percentage of voters who label themselves independent keeps going up. The number who reject both parties more than doubled in recent years. Ticket splitting is commonplace. Surveys show that most voters, even while professing their dismay at Washington gridlock, would actually prefer to have one party in the White House and the other holding the majority in Congress. The parties' grassroots connections, once the key to their political strength, have withered away. In 1993, the then new chairman of the Democratic National Committee, David Wilhelm, complained that he could not even find a list of Democratic County chairmen anywhere in the Democrats' fancy new headquarters in Washington, D.C. Today, the political parties that from the nation's earliest days served as the principal vehicles for collective political action, are mostly relics of a bygone era. Their decline has many causes, all interrelated:

The parties failed to adapt to changes in the nation's social structure. As their baseline constituents, the blue-collar and middle-class white populations, migrated to the suburbs, long-standing political alignments dissolved. The widely dispersed suburbs proved less hospitable to party machines than the closely packed cities. And in the cities, the party organizations that once worked effectively to promote the interests of white Protestants, Catholics, and Jews did not work at all for the people of color who took their place in the old neighborhoods. Racial antagonisms undermined the parties' effectiveness. African-Americans, Hispanics, and Asians, the new generation of urban dwellers, tend to shun traditional party participation and be shunned by a political establishment that has been slow to make room for them. Similarly, with employment in the service economy outpacing jobs in industrial production, the blue-collar labor unions, once the bastion of Democratic Party politics, lost most of their political clout. And with the large-scale decline in family farms, the sources of much of the Republicans' traditional strength in the Midwest and rural areas have dried up as well.

The mass media, especially television, usurped the job the parties traditionally performed in reaching out to the people. Radio and television, along with mass circulation newspapers and magazines, provide the nation with most of its political information as well as public entertainment. Television, a medium of hits and stars, turns political personalities into national celebrities comparable to entertainment stars. Voters no longer have to rely on the parties to signal who stands for what and to tell them what they should be for or against. And people no longer look to the parties to provide them with parades, marching bands, and Thanksgiving turkeys. Nor do the parties offer their constituents soapboxes on which to air their views. Television and talk radio have taken on that job.

Today, computers make it possible for even little known aspirants for political office and for special-interest groups of every variety to reach out to citizens directly, avoiding party channels altogether. Presidential candidates from Jimmy Carter to Ronald Reagan to Ross Perot have been able to organize their own personal political networks that do not depend on the party machines. Voter lists, once the prized possession of the local parties and their key source of political power and control, are now available to anyone with a Xerox machine and a personal computer. Professional pollsters have displaced precinct party chairmen as the principal source of information about what the public is thinking. Politics has turned from "labor intensive work," requiring lots of shoe leather and handshaking, to "capital intensive work," requiring lots of money for advertising and direct mail. Money and media have replaced personal contact as the chief energizing sources of politics.

A new industry of freelance professionals has eliminated the need to rely on party organizations to provide the expertise and leadership for political campaigning. Campaign managers, direct mail experts, media specialists, advertising and promotion professionals, and political fund-raisers are readily available for hire by any political aspirant with the money to pay for them. Candidates recruit their own teams of political professionals and put together their own personal organizations that no longer depend on help from party headquarters or the blessing of its hierarchy. And with elected officials no longer so beholden to the parties for money, campaign services, patronage, ideas, or their nomination or election, their fealty to party leaders has all but eroded.

Also not to be underestimated, the advent of the automobile, radio and television, movies, and professional baseball, football, basketball, and hockey have displaced the political parties as chief suppliers of the kinds of entertainment that gather crowds together. And with the rise of the welfare state, government itself displaced the parties' welfare role in providing jobs, turkeys, food baskets, relief money, housing, and health care support to the needy. Party politics, once deeply integrated into virtually every aspect of ordinary people's lives, has grown marginal.

A century-long succession of political reforms also gradually weakened the parties' hold on their constituents. The continuing expansion of the civil service merit system, starting with the federal government and slowly gaining strength at state, county, and city levels, replaced the spoils system and cut sharply into political patronage. The waves of reform all but wiped out the parties' ability to supply jobs to and do favors for the political faithful. Patronage and voting reforms limited what the parties could do for their ordinary constituents but not what they could do for their fat-cat contributors. In response, the parties, in effect, cut back on their retail services to constituents and expanded their wholesale efforts to raise campaign money from special interests. Also, extending the franchise to women, people of color, and eighteen-year-olds opened up the process to new constituents with new agendas that further diluted the parties' power.

In the 1970s, a new wave of populist reforms, introduced in the wake of the civil rights movement and Vietnam War protests, strengthened the direct primary system, further weakening the role of the parties in choosing their own candidates.

In 1974, presidential campaign finance reforms gave candidates federal matching funds for the first time and put a ceiling on contributions that could be accepted from individuals and corporations. This eroded even further the parties' increasingly tenuous hold on the election process. Federal matching funds go directly to the candidates rather than to the parties. And political action committees (PACs) that sidestep the limits placed on political donations by pooling individual campaign contributions, give most of their money directly to candidates instead of to the parties. PACs contribute to candidates rather than to parties because it is the winning candidates, not the parties, who vote on the laws that PACs want to see passed or defeated. With the breakdown in party discipline, office holders no longer take their cues from party leaders as consistently as they formerly did. As a result, the balance of financial power in politics has shifted from the parties to individual special interests, PAC contributors, and the candidates themselves.

Single-issue, factional politics has split apart the parties' traditional coalitions. A new set of highly emotional personal issues that center on race, gender, abortion, and similar controversies has in many cases superseded traditional bread-and-butter political issues and the political coalitions that have formed around them. These values issues have fragmented long-standing party alliances and crossed traditional geographical and economic boundary lines. In 1985, a survey found that 45 percent of Americans believe their political interests are better served by "organized interest groups" than by the traditional political parties. Since that time, the number surely has increased.

The year 1992, a presidential election year that was supposed to be dominated by economic concerns involving jobs, affordable health care, government spending, and taxes, had an added starter, the unlikely issue of abortion and freedom of choice. Despite the fact that the president plays only a marginal role in dealing with abortion matters, one study of the 1992 presidential election found that the question of choice had a stronger influence among voters "than any other issue, including the state of the economy." Political scientists, analyzing the role of social issues in presidential politics, predict that future campaigns "are likely to be dominated by conflict between the growing evangelical and secular sectors, with regular churchgoers squaring off against those who profess no religious faith or skip religious services."

In the 1994 bi-election year, prayer in the schools, so-called family values, and similar issues came to the forefront. In numerous races, evangelical voters were of decisive influence in deciding the outcome. Such ideological conflicts overlap traditional party lines and erode traditional party loyalties. When politics is dominated by interest groups, political scientist V. O. Key, Jr., observed, these groups "seek to attain the adoption of those policies of particular interest to them; they do not . . . campaign for control of responsibility of the government as a whole. Their work goes on regardless of which party is in power. . . . Theirs is a politics of principle," leaving little room for coalitions, compromise, or political accommodation.

Such single-minded dedication to specific causes produces confrontation politics. By definition, single-issue-movement groups, whether dedicated to civil rights, women's rights, abortion rights, or senior citizens' rights, are not primarily in the business of building coalitions for lasting political power as the parties have to be. They have little interest in, or reason to, compromise. Lacking a long-term commitment to anything but their own cause, these groups do not devote themselves to carrying out the continuing work of politics in the general interest.

Periodically in the past, passionate single causes have played a pivotal role in American politics. Indeed, more than a century ago, Tocqueville commented on the unique predisposition of Americans to form voluntary associations whenever they feel strongly enough about a particular cause to want to get the government to do something about it. Abolition, prohibition, women's suffrage, the various "Know-nothing" anti-immigration movements, the civil rights movement, and the anti–Vietnam War movement—each has had a strong impact on the nation's political history. The difference now, however, lies in the greatly increased ease and speed with which individual interest groups can mobilize public support nationwide (and even worldwide) by means of radio and television, interactive telecommunications, computer networks, and modems. With traditional political party loyalty growing weaker, powerful ad hoc interest groups and single-issue political movements have shown they can unravel the balance of political priorities that, taken together, comprise the patchwork quilt of the national interest.

The debate over health care reform in 1993–94 offers a vivid example. To be sure, the Clinton proposal was a complicated mess, but "The system was simply overwhelmed by millions of dollars worth of lobbying, polling and advertising . . . that produced . . . an exhausting paralysis . . . ," *The New York Times* reported. "Modern politics exalts the grass roots, so interest groups spent $50 million in advertising to cultivate 'grass roots' sentiment. . . . Modern politics also reveres the lawmaker who is quiveringly sensitive to the folks back home, so the best-organized interests . . . have become as powerful as any committee chairmen searching for a compromise. . . . Rather than educate our constituents about the tough choices, too many politicians in both parties are once again pandering to them. . . . More than anything, modern politics never goes off campaign footing." As the grassroots members of the public have become the battleground for political action, the traditional political parties have been diminished and undercut.

DECLINE OF THE CONVENTIONS

On an altogether different level, the rise and fall of the great presidential nominating conventions offer a useful illustration of the nation's continuing trend toward direct citizen participation in the political process—toward direct democracy—and away from the intermediated political system favored by the constitutional Founders. That trend, which has been going on for over a century, has accelerated in recent years, propelled by the extraordinary force of television.

In the early years of the republic, the congressional leaders of each party chose their presidential nominees in closed caucuses. The existence of political parties themselves was something of a populist departure from the original intention of the constitutional drafters, who considered parties dangerously divisive to a stable republic. Starting in the 1830s, the closed congressional caucuses came under attack during the populist tide of President Andrew Jackson's administration. The Jacksonians viewed "King Caucus" as an elitist, undemocratic institution. The caucuses were swept away and replaced by open nominating conventions, in which rank-and-file party members, elected as delegates from each state, would gather to vote for each party's presidential and vice presidential nominees. By 1840, the national party conventions had become the established way to choose the presidential nominee and define the parties' political platforms for the following year's presidential election.

For more than 130 years, the national conventions carried on that role, until 1972, when a series of modern democratic reforms took the job of nominating presidential nominees out of the hands of party delegates and gave it directly to party voters in state primaries and caucuses. For the past two decades, the national party conventions have been reduced to rubber-stamping the primaries' popular choices. The conventions have become nothing but cheerleading rallies for the presidential campaign ahead, their substantive political role all but eliminated. Populist political reform, made possible by television and jet planes, deposed "King Convention," much as "King Caucus" had been deposed more than a century earlier. The nominating process, following the long march of American politics toward direct public participation, has moved from appointment by congressional leaders in closed caucuses, to vote by party delegates in open conventions, to balloting by the people at large in statewide direct primaries and caucuses. Indeed, in some states voters need not be registered with the party in whose primary they choose to vote.

Television network news was born at the national nominating conventions, and in the end it was television network news that devoured them. The first news events ever televised by the networks were the Democratic and Republican Party conventions in Philadelphia during the fiercely hot summer of 1948. The two political conventions were ideally suited for television and its big and unwieldy black-and-white cameras in the medium's early days: In 1948, both parties scheduled their conventions in the same hall in the same city. The action took place in a confined space that the cameras could cover. The conventions were filled with colorful and important personalities and celebrities from every state in the union. The fact that influential political figures, responsible for giving out then-new television station licenses, were there did not hurt, either. The conventions featured a well-defined schedule of events so that the networks' coverage could be planned. Yet the sessions themselves, at least in those days, were always unpredictable and often filled with suspense. The conventions had winners and losers, as in championship sporting events, and the results were of great interest to the entire nation. The conventions afforded a relatively cheap way for the networks to fill many hours and attract the public's attention to the already fast-growing new medium.

The television networks were barely seven weeks old when they inaugurated their first convention coverage. Television networking had been born on May 1, 1948, when AT&T launched regular commercial transmission of television pictures to nine eastern cities via a coaxial cable that could carry moving pictures as well as sound. At the time, NBC had no television news department of its own. It handed off the job of televising the conventions to a crew from *Life* magazine with absolutely no experience in either radio or television.

The very idea that members of the audience outside the convention hall, at least those few who had access to television sets, could see the convention proceedings live generated huge interest even in those parts of the country the television networks did not yet reach. Television held the promise, eventually, of informing, educating, and engaging the entire electorate in unprecedented ways. Recognizing the enormous political potential of that expensive novelty, both Democrats and Republicans had picked Philadelphia as the site of their 1948 conventions because Philadelphia was strategically located on the north-south route of AT&T's new coaxial cable that linked nine cities from Boston to Richmond. During that summer of 1948, the emerging medium made its first impact on American politics, an impact that was to transform politics forever.

Both party conventions that year provided the thousands of viewers who watched at home, in bars, and peering through store windows, with exciting, suspense-filled political spectacles. The Republicans were heavily favored to win the election in November. Harry S. Truman, the Democratic incumbent who ascended to the White House upon the death of Franklin D. Roosevelt, was highly unpopular. A failed haberdasher before going into politics, Truman seemed overwhelmed by the job.

The Republicans took three tense roll call ballots before they finally nominated their presidential candidate, Governor Thomas E. Dewey of New York. Dewey overcame two stubborn Republican rivals, Senator Robert A. Taft of Ohio and Governor Harold E. Stassen of Minnesota. The long, hard battle for the nomination, fought in full view of the television cameras, was exciting to watch. It was the last time the party took more than a single ballot to select its presidential nominee. The Democrats, too, put on quite a show despite the fact that the incumbent Truman was heavily favored to win the nomination. At one point in the proceedings, after a bruising confrontation over the introduction of a civil rights plank, half the Alabama delegation angrily rose from their seats and stalked out of the convention hall. Eventually, Truman beat back insurrections on both the left and right— from southern white supremacists on one hand and progressive northern liberals on the other. The gavel-to-gavel coverage made great television.

Following the rowdy Philadelphia convention, Truman launched a tireless cross-country campaign, taking his case directly to the people and blaming all of his problems on the "do-nothing" Republican Congress. His feisty, earthy, man-of-the-people character, by contrast with the bland and cautious front-running Dewey, played well on the campaign trail and proved to be an especially big hit on the new medium of television. Despite southern defections on the right to the new Dixiecrat Party, headed by white supremacist governor J. Strom Thurmond of South Carolina, and a Progressive Party challenge on the left from FDR's former Vice President Henry A. Wallace, Truman pulled off the greatest election upset in United States history. Television launched its maiden political voyage with a "miracle" election in which "every expert had been proven wrong. . . . The people had made fools of those supposedly in the know." Playing to an ever-widening audience at home, television gave the political parties new access to millions of ordinary people who had ringside seats to the colorful, unpredictable, high-stakes game of presidential politics.

By the time of the next political conventions in 1952, the network television cameras were sending their live signals clear across the nation. The contest for the Democratic nomination was fought between the first national political celebrity to be created by television, Senator Estes Kefauver of Tennessee, and the clear choice of the party's political bosses, governor Adlai Stevenson of Illinois. The once-obscure Kefauver had gained instant nationwide name recognition by presiding over a series of sensational Senate hearings on organized crime, televised live on network television. The televised hearings starred the legendary "mob boss" and "gambling czar" Frank Costello and a colorful procession of Costello's unsavory underworld colleagues. Barred from showing Costello's face during his testimony, the live TV cameras focused on the mobster's nervous hands. Television audiences were mesmerized, and chairman Kefauver emerged as a sudden presidential contender, the first but certainly not the last political figure to jump from local obscurity to nationwide fame thanks to the new electronic medium.

Running a man-of-the-people campaign for the Democratic presidential nomination and wearing a Daniel Boone–style coonskin cap, Tennessee's Kefauver entered twelve of the thirteen state primaries and won all but one. Truman had withdrawn from the '52 race not long after Kefauver scored an easy victory in the New Hampshire primary against the president's designated stand-in on the ballot. The New Hampshire primary was the first ever to be covered live by television. Overwhelming as were Kefauver's triumphs in the field, however, the state primaries did not decide presidential nominations; they were merely the preliminary rounds of the presidential contest. There were too few of them to give any presidential candidate a majority of the delegates. The real decisions were still made at the national nominating conventions. And in 1952, the Democratic convention was run by powerful big-city-machine bosses who disliked and certainly distrusted the independent-minded, populist challenger from Tennessee. The party bosses did what they had always done, which would be unthinkable today. Disregarding the overwhelming primary results, they lined up the majority of delegates in support of Illinois's Stevenson, the candidate backed by Chicago boss Colonel Jake Arvey. Kefauver, the television maverick, did not go down without a fight, however. It took fifteen hours and three roll call ballots before Stevenson finally beat back the Kefauver challenge, while the nationwide television audience looked on. It was the last national convention that required more than a single ballot to nominate a presidential candidate.

In the November election, General Dwight D. Eisenhower, the hero of World War II who, although not especially articulate, had great personal warmth and charm on television, overwhelmed the eloquent and urbane Stevenson. The 1952 presidential campaign made history of sorts by introducing the first political television commercials to sell a presidential candidate. Much to Eisenhower's own discomfort, his campaign commercials were produced by a packaged-goods advertising agency, Ted Bates, that was best known for its hard-sell toothpaste, headache remedies, and razor blade advertising. By 1956, four years later, both parties had become so attuned to television that they geared their entire national conventions as much to the viewers at home as to the delegates in the hall, even though it was the delegates who were supposed to carry out the convention's business. In future years, the balance would swing entirely toward the television audience at home, while the convention delegates in the hall served as mere appendages, in effect living background props for the live television cameras. In 1988, while surveying the set-up in the halls for the Democrats in Atlanta and Republicans in New Orleans, I was struck by how the convention's designers had crowded all the delegate seats together in the least desirable locations, at the sides and back of the auditorium. The television cameras and anchor booths, by contrast, commanded the front-row-center positions that had unobstructed line-of-sight views of the convention rostrum and platform. The only way most of the voting delegates in the hall could see what was happening at their own convention was to watch the flickering images on huge television screens hung high over their heads, the modern-day equivalent of the shadows on the cave wall in Plato's *Republic*. The truth was that by 1988 the television audience had entirely replaced the convention delegates as the focus of attention.

In 1956, it hadn't yet reached that stage but that process had already begun. More than one hundred million Americans in some thirty-seven million homes, a 100 percent increase over 1952, tuned in to nine days of live gavel-to-gavel coverage of what CBS called the "Blue Conventions." The conventions were given that name not because their political dialogue was so raw, but because the bulky black-and-white network television cameras worked best against a blue background. As a result, convention officials colored everything in the hall, from delegates' shirts to podium surfaces, in pale shades of blue. Convention delegates were instructed not to wear white shirts or white dresses, or to show white handkerchiefs in the breast pockets of dark jackets. On television, "Tans, light blues and grays will reproduce in a white more flattering than white itself," their instructions said. Delegates in the convention hall, eager to look their best on the television sets back home, bought politically correct blue shirts to comply with the proper television dress code. Even the conventions' traditional state signs—waved by delegate chairmen on the floor when it was their turn to vote or when they tried to attract the chairman's attention for a motion from the floor—were redesigned from horizontal to vertical, so they could be read on the nation's small television screens.

Out of deference to the needs of network television, both parties held stopwatches to their floor demonstrations and called time on the speakers at the rostrum—an unprecedented move. In 1956, the contrast from conduct at all previous conventions was startling. *Time* magazine noted, "TV's impact on the Convention was emphasized from the start when Paul Butler [the Democratic Party chairman] surprised everybody by banging the gavel on time." The Democratic convention's crusty permanent chairman, House Speaker Sam Rayburn of Texas, announced that all demonstrations would be restricted to no more than twenty-five minutes each and seconding speeches to five minutes. And for the first time in the long history of those unruly events, the time limits were actually enforced.

The Republicans, who in 1956 planned to virtually crown the enormously popular Eisenhower for a second term, designed their convention with no less an eye to the television cameras than the Democrats. With no suspense about the outcome to enliven the potentially dull political proceedings, the Republicans embraced the popular world of entertainment. They introduced Broadway and Hollywood stars, name bands, choral groups, ballet dancers, and live background music to keep the restless television audience tuned in. Showbiz displaced politics as Ethel Merman, Irene Dunne, Patrice Munsel, George Murphy, and Robert Montgomery performed on the convention stage for television audiences across the country. Television's conquest of the political conventions was proceeding at blitzkrieg pace and soon would strip them of the very qualities of unpredictability and political significance that had made them so appealing and desirable in the first place.

In 1960, John F. Kennedy, the handsome and rich young senator from Massachusetts with presidential aspirations, broke new political ground by campaigning in state primaries to demonstrate that his Catholic faith would not be a deterrent to election to the White House, as it had been for New York governor Al Smith in 1928. The turning point for Kennedy came in the heretofore comparatively insignificant primary of Protestant West Virginia, where Kennedy's Catholicism had become the dominant and most highly publicized issue. After a tough statewide campaign, Kennedy defeated veteran Minnesota Senator Hubert H. Humphrey, the candidate favored by most Democratic Party leaders at the time. Television had brought intense national visibility to the West Virginia primary and Humphrey's loss there killed his presidential aspirations, at least for 1960.

West Virginia's Democratic primary battle foreshadowed the presidential nomination wars to come, in which the states' primaries and caucuses would displace the public's national conventions in deciding the nominations for president. In 1960, sixteen state primaries chose just 27 percent of the Democratic delegates to the national convention. Twenty years later the number of primaries doubled to thirty-two, electing 71 percent of the convention delegates. The winner of the majority of delegates at the primaries would gain the presidential nomination well before the convention itself would be gaveled to order.

I attended my own first national party convention in 1960. As a junior advertising copywriter at CBS, I was assigned to go to the Democratic convention in Los Angeles to write the definitive description of how CBS News would cover the great event and beat the other networks in doing so. After Los Angeles, I was to move on to Chicago to do the same for the Republican convention. It was taken for granted that CBS News would score another huge ratings triumph over the other two networks in bringing the nation the biggest live television story of the year, as it had consistently done in the past. Four years earlier, CBS had attracted almost as big a nationwide audience to its convention coverage as the other two networks combined.

For me, it was a plum assignment. The trip to Los Angeles was my first-ever airplane ride, my first time at a political convention, and my first chance to go behind the scenes to watch CBS News cover a breaking event. For CBS, it was a perfect opportunity to trumpet its corporate responsibility and superior dedication to public service. The networks had been tarnished by a series of scandals exposing the fact that their most popular entertainment shows—the big-money quizzes *$64,000 Question, $64,000 Challenge, Twenty-One, Dotto,* and others—had all been rigged; their viewers had been deceived. Along with these damaging revelations, an embarrassing set of criminal investigations involving corruption and payola in radio had been launched. Disc jockeys, program directors, and even broadcast companies had been taking bribes to play records on the air to turn them into best-selling hits. Many of the bribes had been paid by record companies owned by the broadcasters themselves. Congressmen, FCC commissioners, newspaper editorials, and members of the public excoriated CBS and the other networks for their reprehensible practices and shabby standards. It would not be the first time.

In trauma, CBS looked upon its presidential-election year coverage as a timely way to atone for its sins, redeem its reputation, secure its endangered broadcast licenses, and regain the favor of the political community that was so important to the broadcasters. My project, an impressive brochure that would describe CBS's triumphant gavel-to-gavel coverage of the 1960 presidential conventions, would be one step in the campaign to restore CBS's public image.

Landing in Los Angeles after a daylong propeller-driven airplane ride across the country, the city's smog had magically lifted for the

Democrats. The sky was a perfect television blue, the air remarkably pure compared to New York City. Sunny, clean, sprawling, and new, Los Angeles seemed entirely miscast as the convention site for the noisy, old-city, machine-dominated Democrats. No major national political convention had ever been held in California before. Television had made it possible. A new era was at hand.

With the still-popular Eisenhower ending his last term as president, the Democrats felt that their turn in the White House had come. They had led the way to California in the new West, the fastest-growing center of political power and money. Many Democrats favored the glamorous young challenger Kennedy, whose movie-star quality seemed to be straight from Hollywood central casting. But many of the party's older and most loyal supporters backed Stevenson, the eloquent but diffident leader who had run twice against Eisenhower and lost. Stevenson's dedicated supporters thought their man had earned the chance to run against somebody else besides Eisenhower, someone who was bound to be easier to beat. A third Democratic candidate, Senate Majority Leader Lyndon Baines Johnson of Texas, backed by powerful party warhorses, was a major threat in the race, hoping to slip through between the two emotional front-runners.

The political maneuvering, rallies, caucuses, and demonstrations on behalf of the candidates and in support of obscure causes—ranging from vegetarianism to oddball California religious sects—spilled over into downtown Los Angeles in addition to surrounding the convention center itself. For a fascinated, starstruck neophyte, the convention scene was the stuff of real-life drama and suspense. But the high point every day for me was breakfast. Every morning at seven I showed up at the CBS News steak-and-eggs strategy breakfast in a large corner suite in the Biltmore Hotel. There the producers, correspondents, and news executives traded gossip about the previous day's developments, shared their news about what would happen in the day ahead, and organized their assignments while I watched it all and took notes.

Much of their gossip centered on Kennedy's promiscuous and hardly secret sex life, which was reputed to have been carried out in widely dispersed Beverly Hills hideaways. Other talk focused on the candidate's father, Joe Kennedy, and his extraordinary machinations to ensure the nomination for his son. No one would have dreamed of putting such stories on the air. Today, no one would dream of keeping them off the air. Politics then was considered a public business and a candidate's private life was off limits for reporting. The CBS newsmen were serious journalists who, like all journalists in those days, loved to dig up juicy personal gossip but would never think of exposing any of it. Politics was very much of a men's club and private liaisons, drinking problems, and similar personal frailties were considered beyond-the-pale for most newsmen to report or discuss on the air.

In the end, Kennedy won on the convention's first ballot. Stevenson's passionate supporters were no match for the Kennedy family's take-no-prisoners political juggernaut. The real work had been done before the convention began, at the state primaries and caucuses and in the back rooms with the still-powerful big-city bosses.

Unfortunately for my project, the torch was passed in Los Angeles not only to a new generation of political leadership but also to a new generation of network news leadership. The fresh, young NBC News anchor team of Chet Huntley and David Brinkley captured the majority of the nation's viewers, winning the Democratic convention coverage ratings in a big way. The national television audience and the nation's critics found Huntley and Brinkley's witty, wry conversational approach a welcome change from the serious, ponderous, and at times self-important broadcast style of CBS News, which was a holdover from radio days. NBC News won the convention ratings battle for the very first time, and my plum promotional assignment was abruptly canceled. Instead of continuing to Chicago for the Republican convention, I was dispatched back home to CBS in New York. Of more importance in the history of television than my own aborted convention assignment was the fact that in 1960 CBS News ousted Walter Cronkite from the anchor's chair at the Republican convention in Chicago. In a desperate and fruitless effort to stem the Huntley-Brinkley tide, CBS News replaced Cronkite with a news team that included veteran radio correspondent Robert Trout and newcomer Roger Mudd. It was about the only glitch Cronkite suffered in his long and extraordinarily successful news anchoring career.

Two years after the decisive convention triumph of NBC News, I moved to NBC as director of advertising. The 1960s were expansive, golden years for the television networks. NBC and CBS, and in those years a far weaker ABC, dominated what seemed to be virtually every aspect of American life to an unprecedented extent. Day and night, the three networks attracted more than 90 percent of the audience watching television. People spent more time tuning in and talking about the networks' entertainment, news, and sports programs than doing anything else except working and sleeping. The new network season each fall—with enduring stars like Lucille Ball, Ed Sullivan, Jackie Gleason, Sid Caesar, Red Skelton, and Phil Silvers, and enduring hit shows like *Gunsmoke, Bonanza,* and *I Spy*—was by far the most eagerly anticipated national event of the year. The fast-rising new medium of television seemed to be the center of the universe.

In 1960 television was credited with winning the presidential election for Kennedy, whose narrow victory was nailed down by the first-ever televised presidential campaign debate. Substantively, there was little difference between what the two of them said. Those who heard the debate on radio thought Republican candidate Richard M. Nixon carried the day. But by 1960, television had risen to become the supreme arbiter of public life in America and shaper of the nation's opinion. And on television Kennedy was a star without peer. His winning smile, youthful good looks, lively humor, and commanding presence gave him the edge over the wan, withdrawn, awkward, perspiring Nixon.

President Kennedy seemed the ideal leader for the new television era until on Friday afternoon, November 22, 1963, it all came to a sudden, stunning, and shocking end in Dallas. For seventy-two hours the three television networks were the lifeline of the nation. History had never seen anything like it; people were transfixed, immobilized in front of their television sets, watching hour after hour of continuous live network coverage throughout an extended fall weekend of national mourning. NBC's cameras caught the bizarre shooting of the president's assassin, Lee Harvey Oswald, in the crowded basement of the Dallas jail, an event that NBC broadcast live. Americans watched, mesmerized by every detail of the solemn funeral ceremonies in Washington—the black military caisson bearing the president's body to the Capitol; endless lines of mourners waiting to pay their last respects as he lay in the Capitol Rotunda; and finally, on Monday, November 25, the remarkable funeral procession led by a riderless black horse, accompanied by President Kennedy's bereaved family and a stunning procession of world leaders marching to the cadence of muffled drums to the burial ground on a hill in Arlington National Cemetery. The president's murder in the presence of television, one observer commented, was "such an overarching event" that all who witnessed it "remember it more for what it did to [them] than for what happened to him."

The following year, at the Republican National Convention in San Francisco, a powerful new movement emerged that began to redefine the character of modern American politics. A new breed of zealous, ideologically driven young political activists captured the Republican Party. Their archconservative candidate, a comparatively little known senator from Arizona, Barry Goldwater, wrested the nomination from New York Governor Nelson Rockefeller and the old-line, mainstream eastern establishment guardians of the Grand Old Party. It signaled the rise of a new style of modern-day "movement" politics. Goldwater and his fellow populist conservatives spurned the traditional coalition-building of the mainstream parties; theirs was a politics of confrontation rather than accommodation, a politics that in 1964 offered "a choice not an echo." Many influences propelled this trend—fear of the rising clamor for civil rights; fear of the power of world Communism; concern about deteriorating moral standards among the nation's youth; an empty and weakened Republican eastern liberal establishment; the fast-growing new political and financial influence of the western and southwestern United States, with their highly appealing frontier mentality and dedication to individualism; the lack of strong middle-ground Republican candidates; the increasing political assertiveness of conservative evangelical movements. Television's "elite liberal Eastern establishment" was one of its targets. And television, which responds to simple, strong, dramatic confrontation rather than subtle compromise and accommodation, helped to accentuate the controversies and accelerate the trend.

The climactic moment of Goldwater's convention in San Francisco was tellingly captured by the commentary of NBC News anchor David Brinkley on the night of July 15, 1964. Brinkley's remarks also referred to a historic moment in the history of television:

"I have just been informed that this portion of the convention proceedings is being fed to Europe by satellite, and I have been asked to explain what is going on. I am afraid that is beyond my powers. It would be hard to explain it even to an American. As best as I can put it for Europeans, the Goldwater people, like those who will demonstrate later for other candidates, have been waiting for weeks and months and years to display in some outward, tangible way how they feel about their candidate. When, finally, he was formally put in nomination, all of the pent-up energy, exuberance, noise, sighs, slogans, bands and balloons suddenly were released. It is partly political, partly emotional, partly propaganda, partly a social mechanism, partly a carnival, and partly mass hysteria. It can be described as nonsense, and often is—but somehow it works."

Television made Goldwater, advocate of "extremism in the defense of liberty," an instant national figure. To many, he seemed both an aberration in the American political landscape and a threat. In the election, Goldwater was utterly overwhelmed by incumbent President

Lyndon B. Johnson. But the Goldwater campaign foreshadowed the new era of movement politics. (Eight years later, in 1972, came the Democrats' turn. The populist Goldwater Republican revolution on the right was mirrored by a populist Democratic revolution on the left, eventually coalescing around presidential nominee Senator George McGovern of South Dakota.)

During the Democrats' 1968 convention in downtown Chicago, the confluence of the youthful counterculture rebellion with the rising intensity of the anti–Vietnam War protest and the passionate confrontations of the civil rights movement—all seen by millions day after day on television's network news broadcasts—unhinged the Democratic regulars. The new television generation of zealous, ideological, youthful, radical activists played to the television cameras and frightened the nation's viewers. Their sit-ins, protest marches, street demonstrations, mass rallies, political pranks, militant threats, random violence, and general unconventional behavior—captured daily on live television—gave the general public the terrible feeling that the entire nation was falling apart.

On March 31, President Johnson, who only four years earlier had won the most one-sided election victory in the nation's history, suddenly and surprisingly withdrew from the presidential race, discouraged by primary challenges from anti–Vietnam War peace advocates. The strong showing in the New Hampshire primary at the beginning of the year by Senator Eugene McCarthy of Minnesota had prompted Johnson to announce, at the very end of a televised special address to the nation on the Vietnam War, that he would not seek reelection, but devote himself to ending the war.

That spring was marked by unprecedented turbulence and tragedy throughout the country, all vividly brought home to the nation's viewers by the ubiquitous, ever-present television. In April, the Rev. Martin Luther King, Jr., the civil rights leader, was shot to death while standing outside of a motel room in Memphis, Tennessee. His murder triggered vicious race riots in 125 cities. In June, immediately after winning a major victory in the California presidential primary, President Kennedy's younger brother Robert, the junior senator from New York, was assassinated in the kitchen of the Ambassador Hotel in Los Angeles. The United States appeared to be careening out of control and television recorded every conflagration and confrontation without respite night after night.

The Democratic convention in August was held in Chicago, the site of twenty-three of the fifty-six conventions of the two major parties.

Indeed, Chicago had been the host city for two of the most riotous conventions in American history, both Republican—the tumultuous 1860 convention that nominated Abraham Lincoln and the emotion-charged "Bull Moose" convention of 1912 in which Republican William Howard Taft won the nomination over ex-President Theodore Roosevelt, who was seeking to make a political comeback. Roosevelt had won 278 delegates in the 1912 primary contests, compared to Taft's 48, but with Taft forces in control at the convention, Roosevelt's lead went for naught, prompting him to go off and establish the break-away Progressive Party.

However, the Chicago Democratic convention of 1968 was to set new records for tumultuousness. A formidable barbed wire and heavy chain-link fence was installed around Chicago's International Amphitheater to keep out the thousands of young anti–Vietnam War protestors who openly threatened to overrun the proceedings. Downtown, in Chicago's streets, parks, and hotel lobbies, Mayor Richard Daley's club-wielding police, reinforced by helmeted National Guardsmen and federal troops, equipped with tanks, armored vehicles, flamethrowers, and bazookas (which fortunately they never used), waged pitched battles against hordes of youthful protestors. With one eye out for the massed television cameras and the other eye out for the police, the kids went into battle chanting, "The whole world is watching." It was hard to tell which side bore more responsibility for the disorders, the police or the rioters. Meanwhile, inside the Amphitheater on the convention floor, the party regulars nominated Johnson's Vice President, Hubert H. Humphrey, who that year had not entered a single presidential primary, to head the ticket. The delegates selected Humphrey on the first ballot; the last time a presidential nominee would actually be chosen at a national convention. As my predecessor at NBC News, Reuven Frank, wrote about the Chicago rioting many years later, "It would not have happened if television were not there, but all we did was be there."

Like Truman two decades earlier, Humphrey did his best to overcome the severe handicap of a badly split party. But unlike Truman, Humphrey lost the election to the supreme master of political comebacks, the resurgent Richard Nixon. (He did lose by a much closer margin than anybody expected, however: 43.4 percent for Nixon to 42.7 percent for Humphrey, with less than 14 percent going to third party candidate George Wallace of Alabama).

THE DECLINE OF THE CONVENTIONS AND
RISE OF THE PRIMARIES

Immediately following the 1968 election, the call went out to "democratize" the Democrats' nomination process and take control away from the party's professionals who had picked Humphrey, ignoring the results of the primaries. The Democrats created a party reform commission, headed by South Dakota senator George McGovern and Minnesota Congressman Donald Fraser, charged with the job of empowering the party's rank-and-file voters to become the new kingmakers and pick the head of the ticket. The reform commission decreed that primaries and open party caucuses would be the "two acceptable methods by which a state could select its delegates to the national convention." Under the new system, all serious presidential contenders in the future would have to appeal directly to the voters in the state primaries and caucuses in order to win enough delegates to capture the nomination. A Humphrey-type campaign to bypass the primaries and seek the nomination at the convention itself could no longer hope to succeed.

The McGovern-Fraser commission also recommended reforms that would have the Democratic Party's decision making more accurately reflect the percentages of women, African-Americans, Hispanics, Native Americans, and other traditionally under-represented groups in the general population. The Republicans soon followed suit, with democratizing reforms of their own. With power transferred downward to the voters in state primaries and open caucuses, the Democrats and Republican Party professionals, in effect, lost control of their own nomination process.

THE ROLE OF FEDERAL
ELECTION FINANCE REFORM

In 1974, federal election finance reforms helped to accelerate that trend. Presidential aspirants received matching campaign funds. Caps were imposed by law on the amount of money that individuals and corporations could donate to presidential candidates who accept federal matching funds. No longer could the party itself decide which presidential aspirant should receive how much money. Candidates found it necessary to start campaigning for the nomination early, if only to raise enough money to earn the federal match. No longer could the party organization's favorite choice wait until the end of the primary process, relying on a last-minute surge from a few of the regular "fat cat" contributors to put him or her over the top.

The changes have had a profound effect. Democratic dark horses like George McGovern, Jimmy Carter, Michael Dukakis, and Bill Clinton mastered the state-by-state primary process and beat out organization old-timers for the nomination. In California, New York, Illinois, Texas, New Jersey, Pennsylvania, Ohio, and Florida, well financed men and women with little background in party activity, little support from party leaders, and little allegiance to their party's programs have beaten organization favorites in a good many gubernatorial and congressional nomination battles.

Victory has gone to those who can master the intricate courtship of delegates state by state, who can deploy their forces skillfully through the long and complex maze of primaries and caucuses, and who can best turn television and, to a lesser extent, other media to their advantage. It puts a premium on technological and tactical skills; on sheer endurance and perseverance in the field, on the ability to come across on TV and to mobilize grassroots support, at times over the opposition or indifference of political regulars. It is a system that puts less of a premium on the ability to govern, to manage large institutions, and work with powerful political colleagues, than on reaching out to the people-at-large.

The "old politics," as it had been practiced virtually since the country began, has come to the end of the line. Movement politics and interest groups have emerged as the dominant forces of the democratized "new politics." Presidential primaries, once relatively inconsequential political beauty contests and preliminary testing grounds for presidential hopefuls, have been transformed into the decisive battlegrounds in which presidential nominations are won or lost. Since 1972, the shift has been away from "the circus of the conventions," dominated by political bosses and party insiders, to "the traveling road shows" of presidential primaries and media feeding frenzies.

THE RISE OF DEMOCRACY
BY POPULAR INITIATIVE

The astonishing rise in the use of ballot initiatives and the growing popularity of imposing term limits on federal, state, and local elected officials have also accelerated the decline of the "old politics" and ascension of the new. By eroding the powers of legislatures, and even governors, these political innovations (for the United States at least) are fundamentally altering the traditional checks and balances of our representative republic and replacing them with what the California Commission on Campaign Financing described as "a new form of 21st century governance."

The movement for ballot initiatives has cut across party lines. It seeks to put legislative power directly into the hands of the people and circumvent the long-standing institutions of representative government. Term limitations, imposed to convert longtime incumbents into temporary job holders, have a similar effect. They decrease the influence of entrenched political professionals and increase the say of voters at large.

As discussed in Chapter 1, California and a growing number of states and local jurisdictions are introducing a form of direct democracy to our representative republic by greatly expanding the use of ballot initiatives so that ordinary voters can pass or reject laws affecting almost every important political issue. Through the initiatives, the public at large in various states has voted on laws involving gay rights, immigration, insurance, education, taxes, transportation, the environment, term limits, lotteries, gun control, reapportionment, housing, crime, and campaign financing. Twenty-three states and the District of Columbia, as well as hundreds of local jurisdictions, now provide for some form of ballot initiative process. Three more states allow citizens to use popular referenda to petition against measures their legislatures have already passed, in effect giving the people veto power over laws their elected representatives have enacted. As public confidence in representative government's capacity to deal with the nation's problems has eroded, the initiative process has continued to expand at the expense of the legislative process. Originally introduced to this country as a safety valve and remedy of last resort against unresponsive or corrupt state legislatures, ballot initiatives today have far outstripped that original narrowly defined vision. They have become a standard instrument of government that shifts the power to make policy directly to the voters. In California, for example, the use of initiatives has jumped fivefold since the 1960s. Spending on California initiative campaigns has gone up by 1,200 percent since the 1970s.

While designed to give people more direct control over the laws that affect their lives and property, the ballot initiative process has not been immune from serious criticism for its own inadequacies. Complex measures cannot readily be turned into simply stated, understandable ballot blurbs. By shifting the focus to single issues that require only yes or no votes, initiatives undermine the process of compromise and the ability to form political coalitions. They discard the traditional system of checks and balances of representative government. They close out the opportunity for careful deliberation and considered discussion by experts and experienced legislators. They greatly increase the role and influence of the mass media, especially the new electronic media, in translating initiative questions in ways the public can understand. They let through occasionally ill-conceived, rash, or poorly drafted schemes for the general electorate to consider. They leave voters overwhelmed by the number and complexity of measures on the ballot. They offer little protection against heavily financed, one-sided campaigns and special-interest propaganda efforts. And major political issues are often decided on the basis of very low voter participation.

Still, ballot initiatives and referenda, along with growing public support for term limitations (a measure that has won almost every time it has been on the ballot), demonstrate the public's growing dissatisfaction with the "old politics" and its determination to temper representative democracy with direct voter involvement in the major decisions of government. As we have already pointed out, California's Proposition 13, limiting taxes, was instrumental in setting the priorities for the Reagan presidency. Proposition 187, limiting access by illegal aliens to government services, had much the same effect in 1994. In succeeding years, direct public voting on issues, even if only in an advisory capacity to the Congress and president, may well also extend to the federal government.

Next, we consider exactly how such direct voter involvement might work in the electronic republic in the years ahead, how that will affect the responsibilities and relationships of the traditional branches of government, and what kinds of policies and reforms are needed to deal with the problems and opportunities the electronic republic will generate in the next century.

PART TWO

7

THE SHAPE OF THE ELECTRONIC REPUBLIC: THE CITIZENS, THE CONGRESS, THE PRESIDENCY, AND THE JUDICIARY

In February 1994, several thousand parents who teach their children at home rather than send them to school flooded members of Congress with so many letters, faxes, and telephone calls that, according to one report, they "shut down the Capitol switchboard." The home schooling parents, a dedicated, well educated, widely dispersed single-interest group, were fighting an obscure amendment to the president's proposed schools bill. The amendment required that all teachers be certified to teach the subjects to which they were assigned. While that may have sounded reasonable to most professional educators and many parents with kids in school, it would have been an impossible barrier for parents teaching their children at home.

As one observer remarked, "What was ultimately responsible for the explosiveness of the outcry was an awareness that spread like an infectious disease over the computer networks of America." The parents had been alerted to the amendment through postings on their electronic bulletin boards. Since they are so widely dispersed throughout the country, many of them use computer networks to share information and communicate with each other about their dedicated common interest in home schooling. Concerned about the amendment's potential to cripple home teaching, the parents converted their computer networks into political tools, exchanging hundreds of electronic messages a day about what they should do to defeat the amendment and how to go about it. They talked to each other on-line and organized themselves to act both individually and collectively to inundate Congress with their opposition to the amendment. One parent cheerfully notified the others in her group that she had sent off a fax to Rush Limbaugh, who did, in fact, give the issue radio airtime. The result was an electronic communications run on the Capitol. Congress responded by promptly killing the teaching certification amendment, which originally had been expected to pass easily, by a vote of 424 to 1.

Until the past decade, the technology on which democracy had operated for some 2,500 years had not changed much. Even the conversion to voting machines was fundamentally a way to get an honest and more efficient head count and was not essentially different from the longstanding use of the voice vote and paper ballot. Today, however, electronic voting, polling, and opinion registering technologies are changing in radical ways, although most of these technologies still reside somewhere on the periphery of official democratic practice. The new technologies may be gaining dramatically in use and influence, but they have not yet been granted official recognition and standing. Nobody can phone, fax, or punch in his or her vote from the computer at home—yet. But in Oregon, for example, voting at home has become so popular that up to 20 percent of the electorate submit absentee ballots, and the number continues to grow.

We are only just beginning to experience the powerful political impact of the new personal electronic media—the faxed petitions, E-mail lobbies, interactive on-line networks, keypad votes, 900 telephone number polls, the use of modems, telecomputers, and teleprocessors to share information and register views. In kitchens, living rooms, dens, bedrooms, and workplaces throughout the nation, citizens have begun to apply such electronic devices to political purposes, giving those who use them a degree of empowerment they never had before. Digital computer networks provide the means for like-minded folks across America to unite, plan, share information, organize, and plot small political upheavals as the home schooling parents did. Of course, not only individuals but also large, sophisticated, and well-financed interest groups and professional lobbying organizations also have learned how to use these sophisticated new tools of political influence.

Exactly how and when the new technologies will be integrated with the old democratic practices cannot be predicted with precision. But the information and technological revolution is proceeding apace. "The advantages offered by new technologies are exciting and ines-

capable," wrote Xerox chief scientist John Seely Brown. "Designed
with care, they can offer society the opportunity both to engage more
people in democratic practices and to engage people more directly and
in new ways. . . . [I]nformation technologies genuinely offer the chance
to empower people, both in the work place and in society at large,
who for one reason or another have become either effectively disen-
franchised or merely disenchanted."

The key lies in the phrase "designed with care," because the new
technologies also offer political dangers that are many and serious. In
1994, ABC News anchor Ted Koppel echoed the views of many, and
perhaps even most of his colleagues in journalism, politics, and political science by expressing dismay at the likelihood that these technological tools will create an excess of democracy. "The country may be
moving in the direction of a purer democracy than anything the ancient Greeks envisioned," Koppel wrote. "It promises to be a fiasco.
Opinion polls and focus groups are Stone Age implements in the brave
new world of interactivity just down the communications superhighway. Imagine an ongoing electronic plebiscite in which millions of
Americans will be able to express their views on any public issue at
the press of a button. Surely nothing could be a purer expression of
democracy. Yet nothing would have a more paralyzing impact on rep-
resentational government. . . . Now imagine the paralysis that would
be induced if constituencies could be polled instantly by an all-but-
universal interactive system. No more guessing what the voters were
thinking; Presidents and lawmakers would have access to a permanent
electrocardiogram, hooked up to the body politic."

The brave new world of interactivity, whose political consequences
Koppel finds so appalling, is, indeed, just down the communications
superhighway. And for better or worse, the new interactivity brings
enormous political leverage to ordinary citizens at relatively little cost.
With the prospect of a microprocessor, keypad, or telecomputer in the
hands of every voter, we are, as Koppel put it, unquestionably "on
the verge of giving our politicians a devastatingly accurate radar system that will tell them unambiguously just which way the crowd is
milling."

In projecting the changes to be wrought in the electronic republic
and defining what shape the newly emerging political system will take,
the question is not *whether* the new electronic information infrastructure will alter our politics, but *how*. Our political system has shown
itself to be remarkably resilient in accommodating the nation's expanding democratic impulse. The two-hundred-year-old Constitution

has survived vast geographical expansion, multiple population increases, and enormous demographic transformations among the nation's citizens. Crafted by the leaders of a weak, agricultural, infant nation, the Constitution also has withstood the Industrial Revolution, the rise of cities and suburbs, the remarkable technological and cultural explosions of the twentieth century, and the emergence of the United States as the world's superpower. Now it will have to withstand still another great transformation. Assuming the continuing yearning of the American people to govern themselves and assuming, too, continuing dramatic advances in interactive personal telecommunications technologies, we can begin to draw a more or less serviceable portrait of the electronic republic.

KEYPAD DEMOCRACY

Citizens already have many ways to express themselves individually and in groups, to each other and to their elected and appointed officials. At their disposal in the future will be a diverse selection of portable fingertip and voice-activated telecommunications media capable of sending, receiving, storing, and sorting data and motion pictures of all kinds. People will be able to compose and receive instantaneous computer messages, faxes, letters, wires, and videos, and they will have the ability to direct communications, in turn, to just about any individuals or groups they select. Time and distance will be no factor. Using a combination telephone–video screen computer, citizens will be capable of participating in audio- and videophone calls, teleconferences, tele-debates, tele-discussions, tele-forums, and electronic town meetings. They will, of course, continue to have the capacity to phone radio and television talk shows from cars, homes, and workplaces, and talk to vast audiences simultaneously.

Most citizens will gain access to a good deal of information and data in many different formats, which can be retrieved on order or automatically through "smart" television sets programmed to select and store or retrieve any material on any subject. The material they call up may have been tailormade specifically for their age group, sex, race, style of living, educational level, taste, and individual interests. Some of it the users will pay for. Some will be provided free because it is promotional or public service in nature or because it grinds the ax of a particular interest group.

As *The New York Times* described the debut of a new congressional service available over the World Wide Web on the Internet, "A person who taps into the service, called Thomas in honor of Thomas Jefferson, can call up the full text of any bill introduced in Congress since 1992 and will soon be able to get all new issues of the Congressional Record. . . . Type a keyword like 'mother' and a person can get a list of every bill that mentions mothers—and every nonbinding resolution introduced to praise motherhood, condemn welfare mothers, chastise unwed mothers or provide equity for mothers and fathers in divorce cases."

Using computerized lists and on-line networks for different interest groups, individual citizens will also be able to send their own promotional material, propaganda, and publicity of all kinds in all formats to individuals, groups, and political representatives of their own choosing. There will be a continuing flow of audio, video, and written communications, dialogue exchanges, yes/no votes and polls, position papers and programs, interviews, speeches, presentations, and advertisements—all rattling around in cyberspace and all instantly available on command. And day by day, numerous polls and surveys, both official and unofficial, reliable and meretricious, impartial and self-serving, will take the pulse of the public, and continuously tabulate political opinion.

Using a designated personal code—one's Social Security number, citizen registration number, or special-purpose phone number—each citizen's message, vote, or question will be capable of being instantly tabulated and sorted to determine its legitimacy, what sort of interest group or geographic constituency it is part of, and whether enough citizens care sufficiently about a particular matter to be worth paying attention to.

People not only will be able to vote on election day by telecomputer for those who govern them but also will be able to make their views known formally and informally, on a daily basis or even more often if they wish, regarding the politics, laws, agendas, and priorities about which they care the most. By pushing a button, typing on-line, or talking to a computer, they will be able to tell their president, senators, members of Congress, and local leaders what they want them to do and in what priority order. The potential will exist for individual citizens to tap into government on demand, giving them the capacity to take a direct and active role, by electronic means, in shaping public policies and specific laws.

With public opinion of increasing moment, a remarkable variety of techniques, devices, and measures will be employed to determine what citizens think at any particular time on any particular issue. To the telecomputerized citizens of the next century, today's public opinion polling will seem as crude, primitive, and limited as the first Gallup polls in the 1930s seem to us now. Sample groups of citizens will be empaneled to represent the whole or specific parts of the population. These focus groups will be surveyed to express their choices in some depth and dimension, and then monitored to reflect any changes in their views. Electronic juries will be asked to render judgments on individual public questions. Professional polling companies and political consultants already do a good deal of this work. We shall see more of it in the future, carried out in more structured and sophisticated ways.

"Boiler room" organizations, hired by special interests, will seek to manufacture and mobilize "grassroots" opinion and stimulate the outpouring of selected messages and votes—to make sure that particular viewpoints are heard. They do that now. In the next century, it will become a mainstream business. Computerized political advertising, promotion, and marketing campaigns, targeted with high degrees of specificity, will lobby ordinary citizens with as much intensity as legislators, regulators, and public officials are lobbied today, because public opinion—the fourth branch of government—will play an even more pivotal role in major government decisions.

TOWARD PLEBISCITE DEMOCRACY

Back in 1912, the United States Supreme Court put to rest the basic question of whether procedures involving direct democracy were incompatible with our republican form of government and should, therefore, be held unconstitutional. The issue centered on the constitutionality of the newly introduced state ballot initiatives, the populist effort to bring direct democracy to several western states. State ballot initiatives, the procedure's opponents argued, violated the constitutional requirement that elected representatives, not the people at large, make the laws of the land. Upholding the constitutionality of direct initiatives, the Supreme Court concluded that the initiative process simply augmented rather than "eliminated or superseded the republican form of government and the representative processes thought to be central to it." The court was confident that elements of direct democracy can coexist within the representative republic.

Given the accelerated use of statewide and local ballot initiatives and referenda since that time, national advisory plebiscites, initiatives, or referenda, at least on certain major issues, may well be put in place by early in the next century. Polls consistently show a good deal of public support for such measures. If the American people had their way, the federal government would join the rising number of states and local jurisdictions that use ballot initiatives and referenda to impose the public's will directly on government policy. According to Gallup, 70 percent of Americans want to be able to vote directly on key national issues in the belief that national referenda will help overcome gridlock in government and end the corrupting influence of special interests on politics.

In 1993, 80 percent of those surveyed were convinced that the "country needs to make major changes in the way government works." As one observer put it, if the public had its way, "governance in the 1990s [would quickly be] transformed from an exercise in backroom decision making to an up-front, open, 'we-want-in-on-the-decision-making' experience for citizens."

The entire nation may vote not only for term limitations and balanced budgets but also to use citizens' teleprocessors and electronic keypads to bypass, or override the legislative powers of Congress. This is not much different from what already has been put into effect in states like California, Colorado, and Oregon. As it is, many congressmen already have become accustomed to surveying their constituents on tough, controversial matters before casting their own votes. On especially vexing legislation involving, say, new taxes, the decision to go to war, abortion, health care reform, or environmental questions, Congress may consider it prudent to refer the ultimate decisions to an actual vote of the people. The electronic mechanisms that will make such national votes practical throughout the year will soon be in place.

In the 1994 statewide elections, voters decided nearly 150 ballot initiatives. Oregonians alone considered 19 issues, Coloradans 12, and Californians 10, on everything from gay rights issues, to cutting off public benefits for illegal aliens, to term limits.

"The ballot initiative has become a major generator of state policy in California," said the official report of the California Commission on Campaign Financing, *Democracy by Initiative,* issued in 1992. "Although the idea of 'direct democracy' by vote of the people is an ancient one . . . , nowhere has it been applied as rigorously and with such sweeping results as in California today. If California's trends continue to serve as a predictor for the nation's future, then [we] . . . will also begin to see the emergence of 'democracy by initiative' as a new form of twenty-first century governance." In the twenty-first century, as California goes, so may go the nation.

Ballot initiatives and referenda on federal issues could be made determinative, as Ross Perot suggested be done with tax and budget decisions in a 1992 campaign proposal that generated substantial public support. Or federal referenda could be made largely advisory, in effect telling the people's representatives how their constituents think they should vote, a system that actually existed in four states over two hundred years ago. In the decades ahead, the public also may seek the power to veto laws that Congress enacts, thereby enabling the people themselves to overrule any federal measure they do not like. As we have seen, Switzerland, a country often judged to have one of the world's most effective democracies, has long operated with just such a system. In Switzerland, within ninety days after a law has been passed, if thirty thousand voters from at least eight cantons sign a petition requesting that it be put to a popular vote, the law must be brought before all the nation's citizens to be ratified. A majority of those voting can overturn the actions of their own elected representatives. The Swiss have made the initiative and referendum process the preferred method of dealing with national legislation in their country.

From today's perspective, none of these scenarios for the United States is far-fetched. The use of initiatives and referenda in states and localities has jumped fivefold in the last thirty years. Today the initiative has become a primary tool of governance in many states besides California. On the federal level, as far back as the end of the nineteenth century, proposals were offered in Congress to introduce both advisory and binding national referenda and initiatives. Such efforts were endorsed periodically by supporters on the left as well as on the right. In the 1920s, a pacifist backlash against war generated considerable support for a Constitutional amendment—the Ludlow Amendment, named after its congressional sponsor—that would have required a national referendum before the United States could declare war and send troops abroad, unless the U.S. had been invaded. Many prominent college presidents and political scientists as well as a good many newspaper editorials supported the idea. Federal advisory referenda were conducted among the nation's farmers from 1933 to 1936, under the New Deal, to help determine market quotas for commodities. In some regions, farmers still vote on market quotas.

In 1980, Congressman Richard A. Gephardt, a Democrat from Missouri and currently minority leader of the House, proposed that the federal government sponsor a national advisory referendum process in which voters could express their views on three issues every two years. Under the Gephardt plan, the issues would be selected after public hearings and the referendum results would be nonbinding. Issues not subsequently acted upon by Congress would be resubmitted for the voters to decide. A 1987 Gallup survey indicated that Americans approved of the plan by a two-to-one margin.

In the 1980s, Ralph Nader, frustrated by a federal government that he was convinced had lost touch with ordinary citizens, called for national ballot initiatives so that the people's voice could be heard and government officials bypassed. In a rare moment of agreement, that call was echoed on the right by conservative politicians Patrick Buchanan and Jack Kemp and economist Arthur Laffer. They believed that the national popular will was being frustrated by an unresponsive liberal elitist majority in Congress. Surveys continue to show strong popular support for such direct democratic measures.

If computer-driven electronic keypads were put in the hands of every voter, such national referenda would be relatively easy to conduct on a regular basis. Whether or not the nation actually adopts these or similar measures of direct democracy, unofficial instantaneous public opinion polls will continue to be available on demand. The federal government will have no choice but to operate in a political environment of virtual plebiscites, even if such votes are not officially recognized.

In June 1992, the Nova Scotia Liberal Party in Canada experimented with a ballot by telephone to choose its new provincial leader. Despite a computer failure that delayed the phone-in election by two weeks, four times as many people voted in that election as previously had participated. The novelty of the Nova Scotia telephone vote may account for some of the increase. But it does demonstrate that convenient, accessible, easy-to-use electronic technology can attract people back into the political arena.

Cable shopping channels have installed high-speed, large-capacity computerized systems to process millions of viewers' telephone credit card orders. The same or similar technology can be recruited to tabulate votes, process polls, and count the results of initiatives and referenda, dialed in from anywhere. In 1992, CBS News asked viewers to call in their opinions to a special 800 phone number during a prime-time election campaign special, *America on the Line*. Although the audience response during the broadcast could by no means be considered a representative sample, three hundred thousand phone calls were tallied out of millions attempted that could not get through—far *more* calls than could have been processed in the 1988 election, far fewer calls than will be able to be processed by the 1996 election.

The question is not whether the transformation to instant public feedback through electronics is good or bad, or politically desirable or undesirable. Like a force of nature, it is simply the way our political system is heading. The people are being asked to give their own judgment before major governmental decisions are made. Since personal electronic media, the teleprocessors and computerized keypads that register public opinion, are inherently democratic—some fear too democratic—their effect will be to stretch our political system toward more sharing of power, at least by those citizens motivated to participate.

A TREND NOT UNIQUE TO POLITICS

The political world is not alone in undergoing this kind of transformation. As we look ahead to the twenty-first century, much of society is changing in similar fashion, eliminating barriers between races, classes, sexes, age groups, employers, and employees, and altering long-standing lines of authority. Companies are flattening management hierarchies and erasing the operational separation between managers and workers. Today, it is considered smart management style to empower workers to solve production problems on their own and have them help decide how to improve their own productivity. Critical information, which once traveled only from the top down the corporate ladder, increasingly is moving from the bottom up as well. Managers long accustomed to making unilateral decisions now ask their customers and their workers to "participate," "collaborate," "consult," "cooperate," and "advise" in designing and marketing better products and services. Participatory management, like participatory politics, appears to be the wave of the future.

Even elements of the press are going through a similar transformation. A 1994 Nieman Foundation conference of leading newsmen and -women, including editors, producers, and members of top management, discussed this theme: "Traditionally, news was what the editor said it was. Now, news is what the audience says it is." A new movement called "public journalism," led by newspapers in Charlotte (North Carolina), Wichita (Kansas), and Columbus (Georgia), is seeking ways to reengage ordinary citizens by, among other things, asking them what they want their newspaper to report. Through reader panels, public meetings, polls, focus groups, and other plebiscitary devices, the *Minneapolis Star Tribune, Wichita Eagle,* and *Charlotte Observer,* among others, are enlisting their communities to help shape the contents of their pages.

Public journalism, according to its champions, is an effort to restore the press's role in stimulating public dialogue and civic participation, a role central to newspapers and periodicals in the eighteenth century, but one that largely disappeared when group-owned publications with audiences in the millions displaced individually owned local newspapers and periodicals with circulations in the hundreds. Too many newspapers have come to resemble the description of ancient Greek choruses: "Old citizens, full of their proverbial wisdom and hopelessness," delivering "stupefied, reverent and somewhat dark attitudes" that nobody heeds. Now, concerned by the steep decline in newspaper readership and increasing competition from all forms of electronic media, many newspapers are reaching out to the public to help them regain a meaningful role in their communities. In a sense, it is a throwback to an earlier day, before journalism became a full-time occupation, when printers filled their periodicals with articles from farmers, lawyers, merchants, and tradesmen in the community who discussed issues, wrote essays, and exchanged views. The Federalist Papers, for example, originated as a series of periodic newspaper articles and essays that sought to analyze and promote the provisions of the new Constitution. The authors were not full-time journalists or professional columnists but leading citizen participants in the Philadelphia constitutional convention—James Madison, Alexander Hamilton, John Adams, and John Jay.

THE DECLINE OF GEOGRAPHY

The declining importance of geography will be another characteristic of the political process in the electronic republic of the twenty-first century, although our political system will continue to be locked into geographical constituencies, which are basic to our constitutional framework. To give citizens a larger voice in the "conversation of governance," some have proposed to expand legislatures as a way to reconnect politicians to their constituencies. A *USA Today* editorial, for example, urged "doubling or tripling the number of representatives."

As Vice President Al Gore pointed out, "today's technology has made possible a global community united by instantaneous information and analysis." He cited the examples of protesters at the Berlin Wall communicating with their followers through CNN and demonstrators at Tiananmen Square who connected with supporters in the United States and with each other by fax machines. But far too little attention has been paid to the tendency of today's technology to fragment the electorate so that it can be reached more effectively with particular appeals crafted for audiences with specific interests. Interest politics has largely replaced sectional politics. New networks and communications media are specialized, stratified, and narrow, carving up the nation and the world into a series of separate, suspicious, and often hostile enclaves. As one journalist remarked, "In the worst case, the global village becomes a global Bosnia, albeit a less violent one."

LOOKING FURTHER
INTO THE CRYSTAL BALL

With the accelerating speed of technological change in communications, in the twenty-first century there will inevitably be a variety of brand new telecommunications experiments in direct democracy, some impossible to foresee from today's perspective. A number of experiments already have been conducted with mixed results. In Hawaii, for example, a "Televote" project in the 1980s sought to get a cross section of citizens to participate in electronic town meetings, read brochures explaining the issues, and then phone in their votes on the issues at hand. (It proved to be virtually impossible to sustain citizen interest, especially after participants realized that their involvement was purely experimental and would have no real effect on policy or laws.) In Reading, Pennsylvania, an experimental interactive cable channel for local politics has for years been putting viewers directly in touch with their legislators.

Alaska has set up a Legislative Teleconferencing Network and Legislative Information Network to stimulate grassroots political participation throughout that state. The legislature takes sworn citizen testimony on the network so that those in remote areas can appear before the Alaska assembly without having to travel to the state capital, which is often inaccessible by road. The network also transmits legislative hearings statewide. Viewers at home receive regular legislative briefings and can hold tele-discussions with their elected representatives in the state capital. Citizens can send what are called officially "Public Opinion Messages" to their legislators through the Alaskan Legislative Information Network.

Other states, from Iowa to North Carolina, are experimenting with similar interactive telecommunications town halls, teleconferences, town meetings, and on-line networks. A group in California is seeking to set up a statewide interactive version of C-SPAN to carry legislative and other governmental deliberations and official events. So far, these efforts typically have involved a very small minority of the population.

Nationally, the White House and many members of congress are accessible on-line through Internet and various commercial computer networks. Locally, Santa Monica, California, has an on-line public access network for residents to communicate directly with city officials and departments. The point of all these changes is that in the future, grassroots democracy will not have to be expressed by citizens pouring out-of-doors, crowding into town squares, or marching on Washington, D.C., state capitols, or city halls. More often, grassroots democracy will be expressed by individuals sitting quietly in their kitchens and living rooms, who are linked electronically to their fellow citizens and elected representatives, and who manipulate wired and wireless computers, television screens, and keypads.

Such a scenario for the future fulfills the prophecy made last century by France's Tocqueville and Britain's Lord Bryce, two of history's most astute observers of the American democratic process: "By whatever political laws men are governed in the age of equality," Tocqueville wrote, "it may be foreseen that faith in public opinion will become for them a species of religion, and the majority its ministering prophet." Similarly, Lord Bryce predicted a time in America when "public opinion would not only reign but govern," and the will of the majority would "become ascertainable at all times, and without the need of its passing through a body of representatives, possibly even without the need of voting machinery at all." For the American people, that time will soon be at hand. For many it is already here.

THE CONGRESS

With the growth of the electronic republic, Congress, too, will change the way it operates in important ways. For two hundred years, members of Congress traveled to the nation's capital in order to govern. By the beginning of the twenty-first century those trips no longer will be nearly as necessary. Senators and representatives will be able to vote, debate, and even participate in caucuses, committee meetings, and hearings without setting foot on the floor of the chamber or in the meeting rooms of the Capitol. With computers, keypads, teleconferences, and videophones keeping them in touch with Washington, they'll find themselves able to spend more time at home, close to their constituents. In what would be truly revolutionary, or at least a throwback to an earlier time, many may even be able to live at home instead of moving to Washington.

As the widespread calls for term limits and balanced budget amendments demonstrate, citizens are seeking to gain more control over their elected officials. Increasingly sophisticated electronics will enable members of Congress, like professionals in most other fields, to do their work and communicate with their colleagues regardless of where they happen to be located physically. Live appearances by means of interactive video technology will be commonplace; a virtual presence in cyberspace will be almost the same as physically being there. Former Education Secretary Lamar Alexander began his campaign for the 1996 Republican presidential nomination in 1994 with a weekly *Republican Neighborhood Meeting,* fed by satellite TV that links GOP activists at two thousand sites. Commented one observer, "Mr. Alexander skillfully blends a message of high-tech and small government . . . , [bringing] government down to human scale." Alexander also campaigned to limit sessions of Congress to six months a year.

By spending more time in their home districts than in the nation's capital, members of Congress can become more accessible to local teachers, barbers, and businessmen than to Washington lobbyists, influence peddlers, and special interest pleaders. Their local media and local constituents will have more opportunities to watch them at work close up. The cozy, closely knit, incestuous political community inside the Washington Beltway no longer will be so cozy and closely knit if legislators go to work by means of long-distance electronic remote control. At least that is one theory.

"The present system evolved over the past 200 years for simple historical reasons," Columbia School of Business professor James Rogers wrote. "Congressmen had to go to the capital. . . . But no one starting a government at the end of the 20th century would devise such a primitive and archaic system. Present and future technology gives us enormous options." Among those options, as we have seen: more direct citizen involvement through advisory or even binding national referenda and ballot initiatives, and petitions to initiate legislation as well as veto laws already passed.

If current trends continue, Congress would evolve into a deliberative body that organizes inquiries and hearings, develops information about issues, and makes decisions after receiving instructions from its constituents. Congressmen will tend to operate more like Lord Burke's once disdained "weather vane," than as a distant and independent body with a mind of its own.

THE PRESIDENCY

Journalists and political observers have noted the arrival of a new style of presidential leadership—"governing in the campaign mode." As two political practitioners commented after President Clinton's first year in office, "Neither the President nor the media seems willing or able to let go of the campaign experience."

To accomplish any significant change, whether in domestic or foreign affairs, presidents must first mount a national campaign designed to gain the support of citizens at large. Recognizing that presidents no longer have the luxury to govern for four years on the strength of one election, *Newsweek* reported that soon after the 1992 election, "Clinton's team plans to turn the White House into a mighty Wurlitzer of political propaganda." Another observer described "a war room . . . in the West Room of the White House with the goal of creating a 'perpetual campaign' to maintain public support for the new President and his programs." Ultimately, the ambitious intentions of the Clintonites were undermined by the ineptness of their execution.

During Clinton's first trip abroad since his election in 1992, *The New York Times*'s R. W. Apple, Jr., analyzed the increasingly influential role that public opinion played in his foreign policy decisions: "Behind all foreign policy there lurks, or ought to lurk, careful political calculation. What the folks at home will not support, the traveling potentate cannot afford to promise." After his first year in office, Clinton himself described the major lesson he learned, "If you want to change the country, you have to have a Congress to help you, but you have to find another way to speak to the country, and I determined that I would try to do that."

Speaking to the country has become the chief task of the nation's chief executive. *The New York Times* columnist William Safire looked back with longing on the days when presidents could invade Caribbean nations like Haiti, Panama, the Dominican Republic, and Grenada without first getting specific permission from the public or even the Congress. No longer can the United States engage in surprise interventions, Safire said nostalgically. Public notification is required in advance.

Despite the modern trappings of the imperial presidency, concern about public opinion—what will the people think?—sharply circumscribes the options of what is supposed to be the most powerful and is certainly the most visible office on earth. In the future, the limits on the presidency will grow stronger not weaker, as every aspect of the president's conduct in office becomes the focus of ever more relentless scrutiny from the incomparably big brotherhood of the entire citizenry. No longer is it a matter merely of courting a few congressional committee chairmen, dealing with influential groups of Washington political power brokers, and wooing the establishment press. The president must go to the people and woo the entire country.

THE CHANGING ROLE
OF POLITICAL LEADERSHIP

To show how much the White House role has changed: In the nineteenth century, presidents considered it unseemly even to take part in controversial policy debates. Explicit presidential discourse on policy issues was exceedingly rare. With the rise of the "plebiscitary" and "rhetorical" presidency of the twentieth century, presidents found it necessary to "speak directly to the people on important matters instead of communicating their views in writing to Congress."

FDR was the first president to address the people directly on any kind of regular basis. He spoke on radio from the White House by means of his famous occasional but relatively infrequent "fireside chats." Eisenhower was the first to allow his presidential press conferences to be on the record, for direct attribution to the president, although actual quotations of what specifically he said, were not permitted. Kennedy was the first to conduct televised press conferences live. The relatively recent accessibility and democratization of the presidency has produced the need for "the kind of leader that nurtures and facilitates public opinion and public involvement—that persuades, teaches, educates." *Newsweek*'s Meg Greenfield called it "facilitating leadership," requiring a delicate balance of appearing to be in charge, while not running too far ahead of where the people are. Now that balancing act has become more precarious than ever.

The Reagan White House was masterful at this kind of effort. Richard Wirthlin, the president's public opinion specialist, conducted continuous polling and held virtually daily consultations with White House officials about how to reach the people and what specifically to say to them. Wirthlin's techniques measured presidential speeches "moment by moment and word by word. The results could identify 'power phrases' and 'resonators,' the lines and tones most effective in altering feelings." Reagan, the president, had a clear and firm conservative antigovernment vision that the people knew and respected, even if many of them did not agree with him. Reagan, the actor, was consummately skillful at using the techniques and mechanics of communication to bring the audience along with him or at least to give him the benefit of the doubt. He and his aides worked long and hard at that part of the job. Cultivating Reagan's direct relationship with the people ranked among their highest priorities.

Modern democracy does not encourage what historian Garry Wills calls "Periclean type of leadership," the leader who dictates to others, who speaks from the pulpit and expects followers to obey. Nor can an effective leader appear to be "mainly a follower doing what the community says when it 'speaks' through elections, through polls, through constituent pressure." Unless the leader demonstrates a core of convictions, as Reagan clearly did, the public views his or her "good followership" as weakness.

Effective political leadership in the electronic republic requires a paradoxical blend of persuasion, conviction, and judgment; an understanding of where one's followers are and how to reach them; an ability to "mobilize others toward a goal shared by leader and followers," and a willingness to compromise if necessary to reach that common goal. Leadership has turned into "a *mutually* determinative activity on the part of the leader and the followers." For both the president and the press, "the landscape of American politics has shifted beneath their feet. Marshall McLuhan was half right. The medium has changed the message but it has also transformed the audience. . . . The 1992 presidential campaign marked the beginning of a new era where the audience calls the shots in the three-cornered conversation between politicians, press and the public." The job of leader in the century ahead, in other words, will be simultaneously to persuade the public, mobilize the public, listen to the public, and follow the public.

The faster information flows, the less people feel the need to rely on proxies to make decisions for them and the more they want to decide for themselves. For most of history, before public opinion about a particular issue could be formed—much less revealed, reported on, and responded to—the time for decision making had passed. On some issues, information took so long to disseminate across the country that the public would have no opinion to express in a timely fashion. As long as communications over long distances were slow and erratic, national political leaders had the leeway to call their own shots and exercise their own discretion largely unencumbered by what the people were thinking. If people did not like the decisions once they finally caught up with them, then they would vote someone else in when election day finally came around.

In 1803, when Napoleon suddenly decided to sell the Louisiana Territory to the United States, President Thomas Jefferson immediately accepted the offer that would more than double the size of the nation. It was a monumental and radical political decision, yet hardly anyone even knew of the transaction at the time. Only long after the deed had been done did Jefferson persuade the U.S. Senate to ratify the treaty that so dramatically changed the face of the nation. In today's world of live communications and television-led democracy, that would be unthinkable. No president would dare take on an initiative of that cost and dimension all by himself, without any prior public discussion, consultation, or approval. As one political mailing piece phrased it recently, the president, for all his apparent power and prestige, has been reduced to being the people's "hired help."

THE COURTS

As the political system grows ever more responsive to majority impulses, and the legislative and executive branches feel increased pressure to bend to the public will, the judiciary remains the branch of government in the best position to serve as a brake upon the people. With television cameras in many courtrooms, the judiciary, like other branches of government, is coming increasingly under public surveillance. But unlike elected officials, federal judges—while not entirely immune from pressures to conform to prevailing sentiment ("th' supreme coort follows th' iliction returns," Finley Peter Dunne's Mr. Dooley insisted)—are not directly accountable to public opinion. Appointed for life, they remain the most insulated of public servants. They can strike down laws and even referenda and ballot initiatives that the majority of the people have approved but that exceed constitutional bounds. (A strong case also can be made that members of the government's permanent professional bureaucracy—its agency and departmental officials, civil servants, and staff aides to elected officials —who operate essentially outside the range of public scrutiny and accountability—have experienced the greatest increase in actual power and influence. In a government with term limits for elected officials, the power of the staffs that stay on will only expand. Paradoxically, in Washington, D.C., the more invisible and anonymous staff members are, the more influential they tend to be.)

True, "a nation that traces power to the people's will does not easily digest the practice of unelected and unaccountable judges' denying the populace what most of them appear to want. . . . A judicial decision striking down a voter effort . . . risks engendering a perception by the public itself that its will has been subverted." Notwithstanding that risk, under the Constitution, the judiciary is in the best position to blow the whistle on runaway majorities. In the electronic republic, it is fair to predict, the courts will need to play an even more pivotal role in checking and balancing the public's will than they ever have done before. The judiciary will continue to have the last word.

A few state constitutions bar their states' judiciary from having any power to review the constitutionality of popular initiatives after the people have spoken. They reason that the people are sovereign, so the people, not the courts, should have the last word. One academic supporter of that view explained why: "While the initiative and referendum may not fit into a given philosopher's model, and . . . may . . . be misused from time to time, one would hope that the courts will not fall prey to the elitist argument that the people do not know what is best for them and therefore need someone else to tell them. . . . If an occasional 'bad' measure is passed, let those who urge less democracy instead use the tools of democracy to convince the people of the 'rightness' of their view."

Most state courts, however, have not hesitated to overturn popular initiatives that the judges find oppress minority rights. In Colorado, for example, a ballot initiative aimed directly at gays and lesbians was declared unconstitutional by the state court in 1993. The court was not fazed by the fact that by striking down the Colorado ballot initiative it was thwarting the will of the majority of Colorado's voters.

In the electronic republic, the judiciary will have the increasingly difficult and sensitive role of protecting the rights of unpopular minorities and thwarting the popular will when it gets out of hand. Under the constitutional system of checks and balances, the courts have the ultimate responsibility to stop any tyrannical exercise of power, even by impassioned majorities of sovereign citizens. In the absence of such court protection, it is unlikely that any barrier would remain to protect unpopular minorities from being trampled on by majorities who believe they are in the best position to know what is in their own best interest.

As a practical matter, it is inherently more difficult to frustrate the expressed will of the people who hold the ultimate power of government, than to overturn a legislative vote or even a presidential decision. Legislatures and presidents speak *for* the people; they are not the people speaking for themselves. Still, the courts cannot escape the increasing burden of providing vigilant judicial review even of those laws that the voters themselves enact under the evolving democratic procedures of the electronic republic. Protecting the essentially anti-majoritarian doctrine of judicial review will become the key to preserving democracy in the electronic republic and preventing it from succumbing eventually to a popular tyranny or demagogic leader.

These assumptions about the future role of citizens and of their changing relationship to the legislative, executive, and judicial branches in the electronic republic raise a host of new issues and at the same time reawaken questions that have faced democratic societies for more than twenty-five hundred years.

8

THE PERILS AND PROMISE
OF THE ELECTRONIC
REPUBLIC

Are ordinary citizens up to the job of making sound judgments about the day-to-day decisions of government, especially when they can do so only by remote control? Should we be optimistic or pessimistic about the prospects for democracy in the electronic republic of the century ahead?

Optimists cheer the potential of the new electronic media to shift citizens from the passive voice of listeners to the active role of participants and decision makers. Many people tout the new interactive telecommunications technologies as instruments that will speed the world on its journey to more freedom for more people, giving members of the public a greater say in the major decisions that affect their lives. Cheap telecommunications and cheap computers, it is said, will convert the United States into a vast electronic agora, "an Athens without slaves." Individual citizens will be empowered to communicate directly with their leaders, express their own opinions, ask their own questions, and ultimately make their own judgments about laws, policies, and affairs of state.

"The push of the new technologies and the network that connects them will be towards individualization and pluralism." The result, the optimists argue, will be an informed, involved community, a shift of power "from mass institutions to individuals"; "a major force for freedom and individuality, culture and morality."

Unquestionably, interactive telecommunications have the potential to bring enormous leverage to the average citizen at relatively little cost, not only "intellectual leverage, social leverage [and] commercial leverage . . . , [but] most important, political leverage." Vice President Al Gore, the Clinton administration's most enthusiastic and most visible protagonist of the New Information Infrastructure, envisions "an America where poor children sit in front of a television tapping information from the best libraries in the country; where physicians examine patients hundreds of miles away; and where everyone calls up a vast array of newspapers, movies and encyclopedias at the flick of a TV controller." Bell Atlantic's Ray Smith sees "a great flowering of intellectual property, a true Renaissance that will unleash the creative energies" of the nation.

THE AMERICAN FAITH IN TECHNOLOGY

But history suggests caution about such promises. Americans have always had exuberant faith in the power of technology to improve society and bolster democracy. They had faith that "the telephone, the telegraph, the proliferation of magazines and the spread of education would all facilitate . . . new participation . . . [and] empower the public." After World War I, the promoters of radio, the public's "music box," promised that it would become a great new bastion of culture and public service. Speaking of the infant medium in 1922, Secretary of Commerce Herbert Hoover said, "It is inconceivable that we should allow so great a possibility for service, for news, for entertainment, for vital commercial purposes, to be drowned in advertising chatter." RCA's General David Sarnoff, the broadcasting pioneer whose company started both NBC and ABC (originally labeled the Red and Blue networks), confidently predicted that radio would bring culture to the masses and inform, educate, and uplift its audience. Arturo Toscanini, the greatest orchestra conductor of his time, was imported from Italy by General Sarnoff to conduct the NBC Symphony Orchestra on the new medium of radio. Could anybody imagine today any American broadcasting company importing a symphony orchestra conductor, much less employing an entire symphony orchestra? Radio was to serve as a massive force for political enlightenment in our democratic society. The reality, however, never came close to living up to the promise.

After World War II, the addition of the FM spectrum to radio was seen as an opportunity to fulfill radio's long overdue promise by unlocking the three-network stranglehold on the nation's audiences. Hundreds of new stations would bring an outpouring of public affairs, education, information, drama, and fine arts, a new surge in creativity and political involvement. It never happened. With the rise of television, the intensified competition brought about by the addition of so many new commercial radio stations served only to accelerate radio's creative decline.

The arrival of television was heralded with even more hyperbolic expectations than radio—"the great radiance in the sky" that would fight the "decisive battle . . . against ignorance, intolerance and indifference." Along with entertainment, television would bring culture, education, and information to the masses. Television, however, followed in the same mainstream entertainment pattern as radio. It seems unbelievable today, but when I joined NBC as director of advertising in 1962, the network had already produced and broadcast fifty-six television grand operas in twelve seasons. In 1963, before the Metropolitan Opera had moved to Lincoln Center, we ran ads promoting the world premiere on NBC Television of a modern opera, *The Labyrinth*, by the celebrated composer Gian Carlo Menotti:

> The place to sit on an opening night at the Metropolitan Opera is a Parterre Box just above the main floor, or the Grand Tier one level higher still, or Bickford's Cafeteria, across the street on Seventh Avenue. At Bickford's 50 to 80 chauffeurs cluster over English muffins and coffee cups until their employers decamp from the Grand Tier and the Parterre Boxes. The chauffeurs know their opera and their composers. They detest Richard Wagner. "Those Germans go on all night." They love Puccini. "He always lets out by 10:30." And they never see either. But the Sunday afternoon of March 3 will make history, for this time, the chauffeurs, like the chic and millions of other Americans, will see the grandest of curtain raisers: the world premiere of an opera. . . . And this time every music lover, every drama lover—not to mention every muffin-cloyed, coffee-sated chauffeur—can have the best seat in the house.

The author of that advertisement was a young *New York Herald Tribune* reporter named Tom Wolfe, who had just come to the city to make his reputation as a writer. Wolfe's copy expressed the vision many of us working in television then shared: to give "every music lover, every drama lover," and everyone else who otherwise never would have had the opportunity, the "best seat in the house" for culturally enriching fare—from Euripides to Eugene O'Neill, from Bach to Balanchine. We had high hopes for television in those early days.

The Labyrinth, produced in the network's studios in Flatbush, Brooklyn, was not an especially memorable work, and as far as I know has not been seen since. But that NBC televised any grand opera at all, let alone fifty-seven of them over twelve seasons, is worth noting today.

Many of television's early broadcasters felt they had both a responsibility and a unique opportunity to use the new electronic medium to elevate as well as to entertain the masses. The question was not whether commercial broadcasters should carry the fine-arts and serious public-affairs programming, but what the ratio of such high-aspiration fare to popular entertainment should be. Today that question is asked no longer. NBC Opera, like Bickford's Cafeteria, the old Metropolitan Opera House, and the *New York Herald Tribune*, is long gone. The fine arts and serious public affairs have all but disappeared from commercial television, and the ideal of a "best seat in the house" for all is dead. Gresham's Law operates in television's expanded channel environment as it does with inflated currency: Bad stuff drives out the good, and the worst drives out the bad, which is more easily enjoyed than the good. The industry is all business and bottom line, with no regard for the audience's cultural or civic improvement. What gets on the air is whatever will attract viewers, which means what provides instant gratification, relaxation, excitement, and escape—the opposite of programming that requires thought, discipline, seriousness of purpose, and educated attention. The single-minded mission of commercial television today is to produce audiences for sale to advertisers of consumer goods and services.

Still, as each expansion of television came on-line, so did exaggerated new expectations for its improvement. After the mid-century, the start of UHF television, followed soon after by cable, promised to fulfill television's remarkable potential to "teach," "illuminate," and "inspire." The availability of so many additional channels would serve democracy's citizenship needs by providing greater choice and diversity of programming. The prediction again was that new cable television networks would be devoted to original public affairs programming, documentaries, politics, and education. Cable would offer a cornucopia of arts performances, serious drama, science, and quality programs for children and the elderly.

Even at PBS, I saw dramatic evidence of the seductive appeal of new television technology, followed by inevitable disappointment and frustration. The nation's poorest states had been among the first to seize upon the new medium's promise as an exciting and powerful teaching machine that could boost their backward school systems by bringing in the nation's master teachers and finest courses. They saw in educational television the opportunity to get a jump start on quality education at comparatively low cost. It would be the shortcut enabling their poorly performing public schools to catch up to the rest of the nation. South Carolina, North Carolina, Mississippi, New Hampshire, and Maine took the lead in making major investments in educational television equipment, state-of-the-art educational broadcast networks, and classroom television sets.

However, it did not take long before classroom closets and school basements in these states and others across the nation were piled high with broken, and barely used television sets and electronic teaching machines. The era of videocassettes and television on demand had not yet arrived. The inflexible central scheduling of over-the-air broadcast transmissions severely limited the usefulness of educational television programs in individual classrooms. Television broadcasting, it turned out, was no substitute for qualified teachers. Educational television proved to be no magic bullet for the nation's poorest school districts. By the time I arrived at PBS in 1976, the promising notion of *educational* television for classroom use had given way to the idea of *public* television to raise the level of programming for children at home, as well as of culture, information, and civic awareness for the nation's adult viewers.

By 1961, the television medium's hoped for "great radiance in the sky" already had turned into what FCC Chairman Minow called the "vast wasteland," with the commercial broadcasters responsible for "the squandering of the public airwaves." Today, Minow calls the medium "a toxic waste dump," reflecting the still greater polluting effect of its multitude of channels. As *New York Times* critic Walter Goodman reminds us, "Commercial television, so brilliant at giving people what they want, has always been cautious, not to say craven, about giving people more than they know they want or stimulating them to want more. Despite the networks' accomplishments and pretensions, even their news departments tend to operate as much along show-business as educational lines. Mass education is not a profit center."

Regarding its political coverage, Goodman concluded, "These are not novel observations; unfortunately, the condition is chronic and, as became painfully evident in the late political season, it is not getting better. A medium that has shown it can bring information and even ideas effectively to millions is reduced every two years to a tool for stirring up emotions and shutting down minds. As they used to say, that is not healthy for living things. Like representative government."

Will the next information revolution, the new, interactive telecommunications explosion, serve us any better? To be sure, the new generation of flexible, individually controlled telecommunications technologies offers new hope for educational improvement. With videocassettes, television screens even now are capable of being programmed by individual teachers in individual classrooms for individual students, rather than from some single, all-powerful network source. And with the arrival of video on demand, there will be even more flexibility, with greater control by the individual teacher or viewer. Moreover, with CD-ROMs and advanced forms of interactive communications, what goes on the screens can be empowering to the viewer—the individual can respond as well as merely watch—providing a potentially valuable and realistic educational tool.

But technology by itself is inherently neither educational nor frivolous, neither empowering nor debilitating, neither uplifting nor degrading, neither democratic nor authoritarian, neither informative nor manipulative. Its mere arrival on the scene will not assure a revolution in education or in politics. Technology by itself will not automatically determine the path of the future or the nature of the political process, although it is bound to have enormous influence on both. Everything depends on how the technology is used, by whom, and for what purpose. The new, individually controlled, computer-driven telecommunications technology is as yet largely untested, at least in the political sphere. But we know a good deal about the performance of the mainstream media.

Notwithstanding the geometric expansion of channel capacity on the information superhighway, quality programming, serious information, and thoughtful content—whether educational, cultural, or political—will not appear spontaneously to fill the available cyberspace. In fact, under the present system, driven largely by commercial market forces, such fare is even more likely to become the exception rather than the rule. As the number of private communications channels continues to expand, the size of the public sphere, the common ground in which all citizens can discuss and debate the public good, has actually shrunk, closed off to everyone except those who can pay for certain channels. "The tendency is definitely away from the broadcast to the narrowcast channels, away from channels underwritten by advertising or government—and therefore seemingly free—to encrypted pay channels and pay programs in which all information is metered and individuals pay for what they receive."

Left entirely to the marketplace, most program choices to be supplied by the next generation of telecommunications media, like the choices supplied by past generations of media, will be characterized by what economists call "excessive sameness." They will concentrate on reaching the eighteen-to-forty-nine-year-old population group, to the exclusion of all others. The emphasis will continue to be on products that gain the widest appeal and most acceptance within this group. There will be an excess of mindless, commercially viable entertainment and a scarcity of challenging, high-minded, quality fare, whether it is radio and television programs, books and magazines, or movies and music. "The present quality of television programs, with dozens of cable networks competing . . . is not noticeably different from when NBC, CBS and ABC had the field to themselves. There is not much reason to expect that it would be different in the 500-channel era," observed sociologist Leo Bogart, one of the nation's most respected media scholars.

The new generation of telecommunications holds enormous potential to serve the needs of democracy. But that potential will not be fulfilled without a determined and coherent national effort, without fundamental decisions about the architecture and uses of the electronic public sphere, and without the funds to make the uses real.

Quality, creativity, and originality do not expand automatically as the number of outlets increases. By favoring frequent viewers and audiences most attractive to advertisers and most likely to watch box office hits, marketplace economics tend to promote homogeneity not diversity.

THE MODERN PESSIMISTS' VIEW

Looking at the content and character of the electronic media today, pessimists see dire consequences for the future of democracy in an electronic republic. They distrust the judgment of the people at large who, as James Madison suggested, are too easily "misled by the artful misrepresentations of interested men" and "overcome by irregular passion." These pessimists conjure up images of rule by an ignorant, ill-informed majority, poorly served by "the degrading effects of the triviality, banality and vulgarity" of television. They worry that people are swayed by the medium's "fleeting, disjointed, visual glimpses of reality, flickering on and off the screen, here today and gone tomorrow." The public at large cannot make sound judgments about the day-to-day decisions of government, argued diplomat George Kennan, because television, the public's main source of information, "cannot consult the rich voice of prior experience, nor can it outline probable consequences, or define alternatives, or express the nuances of the arguments pro and con."

A television-led democracy too readily will sacrifice important long-term interests for highly visible short-term gains, it is argued. Those who experience an event through the television screen may have the illusion of seeing reality and understanding what it is they are seeing. But they cannot gain the depth of knowledge and analysis required for sound and reasoned decision making. Critics who fear the consequences of too much democracy argue that a political system whose decisions require constant consultation with the vast general public is entirely impractical, even with state-of-the-art interactive telecommunications technologies. Whenever the public becomes directly engaged in a major controversial issue, the process of negotiation, compromise, and deliberation—the essence of effective policy making—becomes difficult, if not impossible. Politicians who might otherwise be inclined to settle matters in the back room, fear that the public will see them as weak and hypocritical if they have to compromise their previously stated position in the full glare of public exposure.

Democracy requires more than merely casting yes or no votes with issues as complex as clean air, foreign trade, health care, and immigration policy. Competent decisions require not only thoughtful consideration and thorough deliberation but also complicated maneuvering through many divergent interests. Many people do not even bother to vote. Most cannot be expected to take the time and spend the effort needed to learn about even the most critical issues that directly affect their lives and well-being. As a result, even in the United States, the world's longest-running democracy, the public's judgment is held in comparatively low esteem.

And that is hardly a new view. For most of the nation's history, the democratic ideal of government by all the people was far from accepted wisdom. After the outbreak of the French Revolution, even the word "democracy" itself had "low connotations . . . [that] became distinctly horrible to respectability" and was held in ill repute for decades. Now idealized and revered throughout the world, the idea of democracy remained in disfavor in this country well into the current century, typified by John C. Calhoun's warning on the floor of the Senate, "The government of the uncontrolled numerical majority is but the absolute and despotic form of popular government, just as the uncontrolled will of one man is monarchy."

George Washington, John Adams, Thomas Jefferson, and James Madison "certainly would have been surprised to find their representative republic universally and vociferously hailed as a democracy." The U.S. Constitution's carefully crafted system of checks and balances, with its independent executive, legislative, and judiciary, was designed specifically to protect the nation not only from an autocratic ruler but also from the tyranny and passions of mass democracy, or "mobocracy." Not until World War I—proclaimed by President Woodrow Wilson as the war "to make the world safe for democracy"—did the United States gain general approval and official sanction as a "democracy." Concern that the American republic will be undermined by excessive democracy, then, is deeply rooted in American history.

CONCENTRATION OF MEDIA OWNERSHIP AND CONTROL

If the first major grounds for pessimism about the nation's future as an electronic republic lies in the traditional fear of too much democracy, that an ill-informed, intemperate majority will have too much say, the second major grounds for pessimism seems to be its opposite. Another group of critics fears that the inordinate power of a rich and privileged few—especially those who own the highly concentrated, increasingly influential telecommunications media—can dominate the debate, influence the views of the many, and manipulate public opinion. In place of democratic pluralism, they argue, we have a dominant telecommunications industry, owned and controlled by a shrinking oligopoly of powerful corporate interests, gatekeepers to the flow of ideas and information.

These critics cite the prediction by John Malone, president of Tele-Communications, Inc., the nation's biggest cable operator, that "two or three companies will eventually dominate the delivery of telecommunications services over information superhighways worldwide. The big bubbles get bigger and the little bubbles disappear." Malone is determined to see that TCI, which already owns cable franchises that enter thirteen million homes and also owns major interests in the nation's leading cable program networks, stays on top of the handful of media giants that remain. "Monopoly is a terrible thing, until you have it," media mogul Rupert Murdoch, who should know, has said.

In 1985, I tried to persuade NBC, then owned by RCA, to seek a controlling interest in CNN from Ted Turner, who was experiencing financial difficulties at the time. As president of NBC News, my theory was that owning a twenty-four-hour cable news service would help amortize the huge news gathering costs of NBC News. It may have been bad public policy, but it was good corporate economics. If NBC could not buy into CNN, which had not yet turned a profit, I urged that it start a competitive cable news channel. Instead, Turner sold major interests in his company to his biggest cable distributors, among them TCI and Time Warner. To protect their financial interest in CNN, the major cable operators—that control programming to tens of millions of cable subscribers—refused to let NBC compete with CNN by offering another cable news channel. They were determined to keep CNN as a twenty-four-hour cable news monopoly.

Later, under GE's ownership, NBC bought the original twenty-four-hour cable Financial News Network (FNN) and immediately folded it so that FNN would not compete against CNBC, its own struggling financial and consumer news channel. Cable operators with a financial interest in CNN had forced NBC to pledge that CNBC would not expand into a general news channel. As NBC president Wright later testified, "A number of large Multiple System Owners [MSOs, the big cable operators] insisted as a condition of carriage that CNBC not become a general news service in direct competition with CNN, which is owned in part by TCI, Time Warner, Viacom, and other MSOs."

In 1993, in the context of a cable retransmission consent situation, CBS tried to get cable operators to accept a competing news channel and also ran into a stone wall. Rupert Murdoch, chairman of the News Corporation, said that he "would have liked to start a news channel but [TCI president John] Malone and [Time Warner chairman] Gerald Levin would not give me the time of day." In an interview in *Broadcasting & Cable,* Murdoch said, "There are at least four companies, perhaps five, that would like to start a 24-hour news channel. The only one that's made a serious effort has been CNBC . . . , but it had to limit itself to business news. . . . But so long as they can't be sure of distribution, they're never going to get the chief executives or the chairmen of those [cable] companies to take the risk and make that investment."

The public was deprived of the opportunity to decide which cable news service (and business news service) it wanted to watch. Imagine if the broadcasting industry or the newspaper or newsmagazine industries were structured so that only one network, one newspaper, or one magazine were allowed to provide a nationwide news service. As former FCC general counsel Henry Geller wrote, "The cable model is a First Amendment horror story: the underlying premise of the amendment is that the American people shall receive information from sources as diverse as possible, yet . . . cable . . . restricts the American people to a single 24-hour news channel."

Overseas, far worse horror stories involving concentrated media ownership can be found. In Italy and Brazil, media empire builders Berlusconi and Roberto Marinho have not hesitated to use their powerful, privately owned media outlets that blanket their respective nations to promote their own personal political ends and drown out political opposition.

Most of what Americans see and hear on television and in newspapers and newsmagazines every day already has come under the control of a few large global companies. And some of the biggest of those—such as GE under Chairman Welch and the News Corp. under Murdoch—are hardly renowned for their dedication to public service, their commitment to quality journalism, or their interest in reportorial fairness and integrity. GE, for one, a company with which I have had personal experience, has a continuing recent history of major scandals involving virtually every major division, including NBC News. Faking a fire in a pickup truck to enhance a news story was by no means an isolated incident. At one of our first meetings, Welch asked why NBC News did not boost its revenue by charging authors, book publishers, and movie companies for the interview and feature time they receive on the *Today* show. They should pay for the editorial exposure they now get free, he suggested, a proposal comparable to urging newspapers and newsmagazines to sell their editorial space to anyone willing to pay the price. GE's internal environment is not one that emphasizes public service or community responsibility.

The prospect of so few big interests, including some of such questionable character, possessing so much communications firepower has dire political implications. And the situation grows worse as multimedia mergers, acquisitions, and consolidations continue apace with hardly a raised eyebrow from the Justice Department's antitrust division.

Concentration and globalization are universal features of the world economy today in industries ranging from oil to drugs and from automobiles to soap. Driven by growing capital needs and the belief that market dominance is a prerequisite for worldwide success, companies in every field buy out, merge, consolidate, and cut deals with competitors. Mass communications, one of the world's fastest-growing industries, is in the forefront of the move toward gigantism. Media titans, like News Corp., Sony, Time Warner, Viacom/Paramount, TCI, Bertelsmann, and Matsushita's MCA, become bound together in conglomerates, interlocking partnerships, and oligopolistic agreements. While new channels of communication explode, the number of gatekeepers shrinks at an alarming rate. It is not good for democracy, nor is it necessary.

On October 13, 1993, CNN carried a press conference announcing the biggest corporate merger of all time, the joining together of two regulated monopolies, heretofore supposedly bitter competitors in building the information superhighway. One was a cash-rich not-so-Baby Bell, the other the biggest cable company of them all. Bell Atlantic, whose dial tone reaches eighteen million homes and operates cellular phone systems serving eight hundred thousand customers, agreed to buy mighty TCI for an astonishing thirty-three billion dollars. The phone companies, flush with capital thanks to their protected status as regulated enterprises, have been impatient to expand into the lucrative television business. And cable companies, which have merged with each other to the point that less than ten control the cable franchises that serve the majority of the nation's cable subscribers, are eager to enter the phone business. (The competing industry's grass always looks greener.) So in 1993, Bell Atlantic and TCI decided to form a telecommunications behemoth worth more than the gross national product of most nations. Rather than compete, they would join forces and carve up the market. Befitting the amount of money at stake, CNN, which also had a vested interest of its own in the deal (its parent, Turner Broadcasting, is partly owned by TCI), televised the press conference live to the world—one advantage of having so many media interests tied together under one corporate roof.

Bell Atlantic's Raymond Smith and TCI's Malone promised that their proposed merger of telephones and cable would be "pro-competitive," would hasten the arrival of an exciting new era of interactive telecommunications, and would offer unprecedented personal choice and unlimited communications capacity. Only through such a merger, they insisted, could the United States stay in the forefront of the telecommunications revolution. The wires of the newly combined companies would reach into the homes and workplaces of more than 40 percent of the nation. The merged company would attract the billions of dollars necessary to construct the electronic superhighway. Unless the two prosperous and powerful companies could combine rather than compete, the two corporate leaders said, each would lose out eventually to even bigger and more powerful world competitors; American jobs and business would be on the line.

Watching the press conference on CNN, I realized that this was almost a verbatim replay, on a much bigger scale, of a script I had heard some seven years earlier. The occasion then was a press conference announcing the merger of two other corporate monoliths, GE and RCA. In 1985, GE agreed to buy venerable RCA for about $6.5 billion, a sum that seemed humongous at the time but that turned out to be a steal for GE and a fraction of the price Bell Atlantic was going to pay for TCI. As president of NBC News, I was an officer of RCA's subsidiary NBC, considered to be "the jewel in the RCA crown." My role in that press conference was to confirm, if asked, that GE had given its solemn promise not to interfere with the work of NBC News. The news division was to retain its independence and integrity under the new owner as it had under the old. In view of GE's substantial defense business and other U.S. government contracts, its major financial dealings with good and bad governments throughout the world, its less than pristine environmental record, and its occasional questionable corporate practices, Ralph Nader and other public interest figures had raised questions about GE's fitness to buy RCA and run NBC. They expressed concern about the independence and integrity of NBC News under GE's stewardship.

I had no reason to believe that GE would operate any differently as the owner of NBC than RCA had for decades. NBC was on top and enjoyed record-breaking profits. RCA had kept its hands off. And both GE chairman Welch and RCA chairman Thornton Bradshaw had given me their explicit assurance that GE would do the same (a verbal assurance that, as Sam Goldwyn said in another context, would turn out to be "not worth the paper it was written on"). At the press conference announcing GE's purchase of RCA, Bradshaw and Welch hailed the new corporate union as the only way to restore America's global competitiveness in the vital electronics industry and fight off foreign competitors. A half century earlier, GE, in fact, had owned 50 percent of RCA, but in 1932 it had been forced to divest its majority interest in the promising new radio company under a Justice Department antitrust consent decree. (NBC's familiar trademark, the three-note musical chime G E C, stood for General Electric Co., the only vestige of that original corporate relationship.)

But this was a new era, marked by fear of rising worldwide competition and a disinterest in antitrust enforcement. Only a reunited GE and RCA, Welch and Bradshaw insisted, would have sufficient size, influence, financial resources, and marketing power to return the electronics industry and its lost jobs back home to these shores. The FCC, the Justice Department, and the stockholders of both companies quickly approved the sale to GE.

The ink had barely dried on the deal, however, when GE sold off the RCA name and RCA's entire line of electronics products to a powerful foreign competitor, Thompson, the French electronics company. GE also shed RCA's David Sarnoff electronics research laboratories in Princeton, New Jersey. Notwithstanding the GE and RCA patriotic press conference promises, RCA no longer exists as a full-fledged corporate citizen of the United States. As a result of the merger, this country lost even more of what little was left of its electronics manufacturing industry and most of the jobs that went with it.

GE also quickly sold off NBC Radio and licensed the use of the name, NBC Radio News, to a company that had no connection with NBC News. Protests that it was unethical and journalistically irresponsible for the NBC News name to appear on news reports that NBC News neither produced nor had any editorial control over or responsibility for, fell on deaf ears. In the manufacturing business, brand names were licensed out all the time for big dollars, GE pointed out. NBC News was no different. And finally, despite all the assurances, GE wasted no time making its mark downsizing and downgrading the quality and character of NBC News.

Listening to Smith and Malone seven years later explaining why their thirty-three billion dollar merger was necessary to restore America's global competitiveness, I wondered if the outcome of their deal would be any different from the GE–RCA merger.

Toward the end of the Bell Atlantic–TCI press conference, a minor but, I thought, prophetic glitch marred an otherwise perfectly planned event. The television sound suddenly went dead when out-of-town reporters were given the chance to phone in their questions. The CNN announcer in Atlanta quickly apologized for the technical problem. Bell Atlantic, the telephone company that supplied CNN with the press conference's television feed, could not get the reporters' telephoned questions heard on the air. It was a small but ironic lapse in Bell

Atlantic's plain old telephone service. Might that be an omen of sorts for the brave new telecommunications world to come?

A few months after the press conference the Bell Atlantic–TCI megamerger collapsed. The two companies could not work out the terms of their agreement after all. Their rhetoric about the merger being absolutely essential to usher in the telecommunications revolution, build the information superhighway, and ensure America's world competitiveness was quickly forgotten. The telecommunications revolution continues apace and undoubtedly even more effectively without the megamerger.

In addition to the rising concern over undue media concentration and the drift toward gigantism in the telecommunications industry, how much credence should be given to the spectacular promise of the coming information superhighway and telecommunications revolution? Will life after television—"the telefuture"—truly "renew our entire culture," "promote creativity," overthrow the stultifying influence of mass media, renew the power of individuals, and "enrich and strengthen democracy and capitalism all around the world"?

The multi-billion-dollar investments in developing the new telecommunications landscape are driven by a simple, irresistibly tempting vision—the prospect of converting every home and workplace in the nation into a computerized electronic movie theater; shopping mall; video game arcade; business, information and financial center; and perhaps even gambling casino, run by remote control and open all day long, every day of the week. The information superhighway will not be a freeway but an automated private toll road, traveled mostly by those who can afford to pay the price for the wealth of popular entertainment, information, data, communications, and transaction services it will carry.

Historical precedent is often important and always useful; in that sense, the new information superhighway will have more in common with the interstate railroads of the last century than with the interstate highways of today. The history of the railroads, whose steel tracks first linked the entire nation a little more than a century ago, offers useful precedents—and warnings—for today's high-tech telecommunications, whose digital wide-band transmissions have started to link the entire nation in cyberspace.

THE EARLY RAILROADS:
A CAUTIONARY TALE

In the 1860s, the federal government sought to accelerate development of the West by granting private companies special inducements to build railroad lines to the Pacific. Today, the federal government seeks to accelerate the development of the new frontier of cyberspace by granting private companies special incentives to build an information superhighway system that will interconnect the nation and, indeed, the entire world.

In the midst of the Civil War, Congress awarded lucrative grants to companies that undertook the expensive and highly risky business of constructing railroad lines across the uncharted frontier: "To the Union Pacific Railroad, authorized to build a line from Omaha to the California border, Congress gave a right of way across public domain, all the timber, stone and earth needed to build the railroad, 20 sections of land with every mile of road constructed and credit ranging from $16,000 to $48,000 a mile." The Union Pacific quickly got the job done. Within a few years, railroad lines bridged the nation, moving people and goods into unsettled territories. The railroads overcame what, until then, had been insurmountable barriers of time and distance. Vast new business empires were created and huge fortunes made from the federal government handouts by a few now-legendary railroad tycoons, men like Jay Gould, William H. Vanderbilt, Jay Cooke, Edward H. Harriman, Andrew Carnegie, and J. P. Morgan.

Not content merely with owning railroad lines, the nineteenth-century railroad barons used their monopoly grip on the nation's long distance transportation system as the lever to gain control of the companies that depended on the railroads to carry their products to market. Resorting to railroad price gouging, rate manipulation, preferential treatment, hidden rebates, and raw financial and political leverage, the railroad moguls bullied their way into ownership of steel mills, coal mines, and lumber companies. Stocks were manipulated and political influence bought so blatantly that by the end of the nineteenth century railroad scandals had reached into the office of the vice president and even, it was alleged, into the White House. Out West, all-powerful railroad interests dominated state governments and even owned several state legislatures outright.

As the nineteenth century came to an end, a rising tide of pressure from the railroads' angry customers pushed the federal government into abandoning its pro-railroad, laissez-faire business policy and start regulating the industry. Antitrust laws broke the railroad barons' powerful grip on other industries. The public's resentment against their rapacious competitive practices drew the net of regulation ever tighter. The Interstate Commerce Act of 1887, the Sherman Antitrust Act of 1890, and various railroad regulation laws passed early in the twentieth century authorized direct government intervention to protect the public interest. The Interstate Commerce Commission was formed to see to it that the nation's railroads operated in "the public interest, convenience and necessity," the exact phrase Congress later incorporated into the 1934 Communications Act to regulate the new medium of radio.

Today's telecommunications tycoons, like the railroad barons a century ago, race to take advantage of lucrative government incentives and exemptions—special easements; access to public and private rights-of-way for cable; exclusive pathways to satellites; and a relaxed view of mergers among cable, telephone, computer, and content-producing media companies—to build their broadband superhighways. Like those dominant railroad companies, the biggest telecommunications companies are buying out local systems, merging with the competition, consolidating long-distance lines, staking out their own rights-of-way, and using their distribution clout to gain ownership of the products they carry. The telecommunications moguls are creating the national infrastructure for the next century. And like the railroads, telecommunications companies use the leverage of their delivery systems to gain control of the companies that produce the products and services they carry to market—the movie and television studios, and now the newspaper, book, magazine, and data publishing companies that make the software that will travel on the information superhighway.

In 1992, after a decade of unbridled and unrestricted cable industry growth, Congress responded to the rising tide of customer complaints about price gouging and poor service by cable operators with monopoly franchises, just as a century earlier it had responded to customer complaints about railroad operators with monopoly franchises. The Cable Television Consumer Protection and Competition Act of 1992 put a lid on cable rates and imposed additional restraints on the cable industry, although the march toward gigantism and concentration of ownership and control continues unabated. The question now is will telephone companies, which are entering the video distribution industry, and direct broadcast via satellite companies, as well as other new-comers, cut into the cable business the way trucks and airlines cut into the railroad business?

THE DISTORTING INFLUENCE
OF MONEY AND OTHER CONCERNS

In addition to innate distrust of the public's judgment and rising concern about the concentration of media power, pessimists point to other disturbing trends that threaten to undermine effective democracy in the electronic republic. Such trends include the increasingly distorting influence of money on the public dialogue; the dangerous growth of professionalism in politics and its apparently growing capacity to manage, manipulate, and exploit public opinion; the expanding role of special interest politics, and the "dumbing down," or debasing, of the standards of political information. We mentioned these concerns in previous chapters, will discuss them in more detail in the following pages, and in the concluding two chapters will suggest policies to cope with the problems they raise.

Every discussion of the democratic process either begins or ends on the critical subject of money, "the mother's milk of politics" and perhaps the central problem of American politics today. The demeaning, dismaying, all-consuming, and all too often corrupting pursuit of money for political purposes inherently betrays the basic democratic principle of "one person, one vote." It rewards candidates, usually incumbents, whose disproportionate access to money gives them disproportionate access to voters. And it rewards wealthy special interests whose disproportionate ability to buy their way into the nation's media skews the national debate on most important issues.

The role that money plays in influencing the public dialogue takes many forms. The growing disparity in the quality of information that is available to some citizens compared to others has a corrosive effect on the democratic deliberative process, on people's ability to make sound judgments, and on the public's faith in the integrity of the political system.

The economics of political information—solid, responsible, and meaningful information about significant issues in contrast to gossip, sensationalism, personal scandal, and the like—do not encourage its widespread distribution in the telecommunications marketplace. The commercial potential of major documentaries, serious discussions, public affairs forums, and analytical pieces about important issues is severely limited, especially if the issues involved are intensely controversial.

In economic terms, "political information is often a public or social good . . . which is of significant but small use to many different people [and so] cannot be sold efficiently. . . . Political information of value to millions of citizens, therefore, probably tends to be underproduced, whereas large, organized political actors can find out what they need to know. A large corporation, for example, with extensive resources and a big stake in political action, has a much better chance of learning how a tax bill will affect it than do many unorganized taxpayers with small, diffuse interests that (in the aggregate) add up to a great deal."

Similarly, special interests with large reservoirs of cash can afford to produce political information that serves their own needs, thereby at times tilting the public dialogue in one direction. In the debate over health care reform, the medical insurance, pharmaceutical, and private health care companies, with vital financial stakes in the outcome, spent tens of millions of dollars in six months of intensive advertising, publicity, lobbying, and one-sided promotional materials. It is certainly understandable. But it is only in the public interest if *all* interests can contend on a reasonably competitive basis.

The financial imbalance can take many forms, some comparatively subtle. During my years at PBS, public television could never raise enough underwriting money from independent sources to produce a single major television series devoted to labor and labor unions. It was a high priority of mine not only because of the subject's innate historic importance but also because of the need to balance off PBS programs, such as *Wall Street Week* and *Adam Smith's Money World*, that placed heavy emphasis on business and finance. Not surprisingly, however, no corporate underwriting, station program funds, or endowment money could be found to finance programs that dealt with labor's side of the equation.

The AFL-CIO's public information department offered to put up union money for a television series on the history of labor. But without matching funds from other sources, PBS underwriting rules prohibited accepting union financing, for good reason. No one would trust a documentary series about labor unions paid for entirely by union funds. AFL-CIO leaders accurately pointed out that unless PBS accepted their money, there would be no programs on public television about the labor movement. Certainly no corporate underwriters, whose contributions are key to public television's national program

production, would be likely to put up the funds. As a result, no program series has ever appeared on PBS, or to my knowledge on any
other television network, depicting the history of the labor movement
in the United States. Public television continues to schedule a full com-
plement of programs about business and finance, for which money
readily can be raised, while there continues to be a dearth of programs
about labor.

Especially galling when it comes to money and politics is the fact
that every dollar a corporation spends on political activities, whether
lobbying public officials in pursuit of the company's interests or buying
ads and commercials to sway public opinion, is tax deductible as a
business expense. Individual citizens have no such privilege. Not only
does the public, therefore, in effect subsidize all corporate political
activities but individuals also must bear the full cost of whatever they
spend to pursue their own political interests.

Without unseemly sums of money to buy access to the electorate,
no candidate for important public office can succeed in today's political arena. In the face of the need for such massive expenditures, the
average citizen is overwhelmed. Thus, political action committees, a
principal source of big money financing for election campaigns, have
"replaced political parties as instruments for electing and controlling
Congress." Even as electronic democracy brings the public unprecedented opportunities to participate in the decisions of government,
"the money onslaught of the PAC system" dominates the debate over
vital national issues.

And the price of politics continues to go up. According to the
Federal Election Commission, from 1990 to 1992 congressional campaign spending jumped 52 percent, from $446 million to $678 million.
An average of more than $1 million was spent for each of the 535
members elected to Congress. In 1992, House incumbents held a five-
to-one financial advantage over their challengers in political contributions. Incumbent senators had a three-to-one edge. Of the top forty
House recipients of PAC dollars in 1992, all were incumbents. In the
Senate, thirty-seven of the top forty PAC campaign recipients were
incumbents. In 1992, the presidential candidates alone spent significantly more than a billion dollars in the race for the White House, of
which more than half went to television.

Can the political money monster be tamed? Ballot initiatives and
referenda originally were introduced at the state and local level by populist reformers for the specific purpose of counteracting the corrupting
influence of big money on politics (especially from the railroads). They

reasoned that while individual politicians and government officials could be bought, the entire electorate could not. However, the dramatic increase in state and local ballot initiatives and referenda in recent years, designed to expand the public's role in its own governance and reduce the influence of monied interests, has succeeded only in raising the financial stakes and funneling political funds into different channels.

Numerous studies of the results of ballot initiatives demonstrate the obvious: Money is a big factor in their outcome. According to the California Commission on Campaign Financing, "Money often dominates the initiative process even more than it does the legislative process. Money alone and in sufficient quantities can *qualify* virtually any measure for the ballot. And money in large amounts can frequently *defeat* any measure so long as that measure is outspent by a wide enough margin." Those who spend the most money win most of the time. In this sense, economic disparity overrides political equality in the information sphere; the marketplace of ideas has grown severely skewed.

In California, more is now spent by private interests to influence voters on ballot initiatives (about $250 million in 1988 and 1990) than is spent to lobby state legislators on all the legislation that is before them. Spending on individual state initiatives for gasoline taxes, bottle returns, health care reform, and smoking in public places exceeded tens of millions of dollars. To defeat the 1990 California alcohol tax initiative, the alcohol industry spent thirty-eight million dollars. Citizens' groups, often outspent by twenty to one, have no hope of matching or coming close to offsetting these massive one-sided efforts. In 1980, Chevron, Shell, ARCO, and Mobil outspent supporters of a proposed California oil surtax by a hundred to one. Early polls showed a majority of Californians favored the surtax. By election day, only 44 percent actually voted for it. A study by the Council on Economic Priorities found that in state initiatives, the corporate-backed side almost always outspends its opponents and wins about 80 percent of the time. Another study that charted seventy-two ballot questions from 1976 to 1982 also found that nearly 80 percent of the time the higher-spending side won.

Efforts by states to redress the financial imbalance, reform campaign financing, and restrict what companies can spend in political advocacy have been consistently overturned by the Supreme Court on First Amendment grounds. In *Buckley v Valeo*, the Supreme Court ruled that spending money was tantamount to speech and held that restrictions on campaign expenditures abridged the First Amendment right to communicate one's views to the public. In *National Bank v Bellotti*, the Court invalidated a Massachusetts statute prohibiting corporate contributions to political campaigns involving ballot measures, on the grounds that the First Amendment protects corporate speech as well as individuals' speech.

Oscar H. Gandy, Jr., of the University of Pennsylvania put the issue this way: "[The growing disparities in political financing] cannot help but exacerbate the massive and destructive inequalities that characterize the U.S. political economy as it moves forward into the information age. . . . These inequalities are emerging in an area that is critical to the maintenance of a democratic polity. . . . These inequalities have to do with differential access to information that is necessary for informed decision making."

Even in a telecommunications system with hundreds and eventually thousands of channels, the full and open marketplace of ideas will not automatically come to pass unless the money problem—the scourge of politics—is addressed. And even those who are optimistic about the public's capacity to come to sound political judgments also worry that the increasing imbalance of money in politics "may distort the public's picture of the world and lead its policy preferences astray." As Page and Shapiro found, based on their study of a half century of national public opinion surveys, "On the one hand, we see the American public as substantially capable of rational calculations about the merits of alternative public policies, based on the information that is made available to it. . . . Collective deliberation often works well. Even when it does not, even when public debate consists largely of outrageous nonsense, the public is surprisingly resistant to being fooled— so long as . . . at least some alternative voices [are provided]. On the other hand [when] the information available to the public may sometimes be overwhelmingly false, misleading, or biased . . . and especially when virtually no dissenting . . . voices are heard—the public may be led astray. The rational public can be deceived."

THE PROFESSIONALIZATION OF POLITICS

For the past half century, politics has assumed a largely individualist caste; rather than participating in group events the electorate has become atomized, reacting to images on television and responding to the mass media. Politics changed from a volunteer communitywide effort to a long-distance, full-time business, a costly professional enterprise that depends on hired squads of experts and specialists: campaign managers, advertising agencies, producers, writers, television directors, researchers, media advisers, political consultants, pollsters, advance men and women, paid signature collectors . . .

The Democratic and Republican parties spend millions on "the empty politics of television commercials and media hype," but nothing any longer "on authentic human conversations . . . , talking and listening to people in their communities . . . , hiring organizers to draw people out of their isolation and into representative political organizations." The old-time party machines, for all their bossism and corruption, served as social centers, public forums, and neighborhood gathering places for citizens to listen and talk about matters that concerned them. Once the focal points of political expression, responsible for both teaching and listening to the people, the political parties have turned into little more than highly organized fund-raisers.

Since the first Eisenhower presidential campaign in 1952, when advertising agencies first produced TV commercials and ad campaigns for political clients, political scientists and others have expressed concern for the increasing professionalization of politics. They deplore the expanding use of powerful and sophisticated new techniques of mass persuasion. In recent years, dramatic improvements have been made in methods of tracking and managing public opinion. Every political campaign of any significance today is run by highly specialized experts, full-time political professionals who have the skills, experience, and resources to orchestrate and exploit voters' emotions and influence their views.

THE RISE OF INTEREST POLITICS

The rise of electronic media that make long distances essentially irrelevant, the decline of the parties, and the loss of community identity all have contributed to the growth of single-issue politics—the proliferation of individual groups of citizens who organize to fight for or against a deeply felt political issue. By nature more divisive than communitarian, single-issue groups are dedicated to mobilizing public support for their own particular causes to the exclusion of all other political priorities, no matter how important the others may be to the entire community. The promotion of individual interests rather than shared values can be dangerous in a heterogenous democratic society.

A public fractured by deeply committed ethnic, religious, racial, or other individual interests can find it hard to agree on what is in the general public interest. If individual citizens approach each issue with disparate priorities and no shared sense of community and civic values, democratic consensus and cohesion will be difficult to achieve.

"DUMBING DOWN" THE QUALITY OF INFORMATION

Finally, instead of ascending to a higher, more responsible level of information, the still critically important mainstream media are ratcheting down their standards to the lowest, most instantly accessible level of sensationalism and scandal. For an electronic republic, in which citizens have the capacity to participate directly in making the day-to-day decisions of government, this is an especially precarious and menacing problem.

"Immediate reward news"—stories involving corruption, accidents, disasters, crime, sexual deviance, and social gossip—has largely displaced "delayed reward news"—stories involving critically important economic, social, and political issues. Faking, staging, exaggeration, superficiality, and polarizing controversy increasingly characterize the performance of even network television, still the most influential of all communications media. Traditional values and old-fashioned rules of journalism have been thrown overboard in the competitive race for audiences and commercial success. The trend toward tabloidization and instant popularization has eroded the boundary lines between news and entertainment, objective journalism and advocacy.

Washington Post columnist David Broder defined the problem this way: "Increasingly, it seems, the standards and patterns for the entire media are being dragged down by those with the shoddiest journalistic credentials—or no credentials at all. Radio talk shows, television 'infotainment' programs and supermarket tabloids have won huge audiences and have demonstrated the capacity to 'launch' stories—often of the sleaziest kind—that the mainstream press feels it necessary to follow."

Once known for aspiring to produce solid, responsible popular journalism, network news divisions now turn over the greatest part of their resources to the production of tabloid style prime time magazine shows and "reality TV." In 1993, NBC introduced *I Witness Video*, a weekly series that in its premiere episode showed four real-life slayings lovingly captured on videotape. *Newsweek* called the program "TV's first 'snuff' show." A few months earlier, NBC's *Nightly News with Tom Brokaw* showed the on-camera murder of a Miami woman by her estranged husband, running the "story" without comment or explanation, and leaving even many of the network's local affiliates horrified. NBC News defended its decision to show the murder on the basis that, well, it had the video. In 1994, NBC News introduced a new prime time magazine show, *Now,* the opening teasers for which included things like "Was sex with children part of this group's teachings?" "Herman and Druie Dutton, ages fifteen and twelve, shot and killed their father." "How would you feel about sharing your spouse?" In the course of one week in February 1994, ABC's Diane Sawyer interviewed mass murderer Charles Manson and two of the women who once killed for him; NBC's Stone Phillips interviewed mass murderer Jeffrey Dahmer, and CBS's *48 Hours* profiled another serial killer.

Americans spend hour after hour watching the most appalling stuff and at the same time bitterly resent television for bringing it to them in such overwhelming doses. Most Americans think television news sensationalizes violent crime; 80 percent are convinced that television is harmful to society and especially to children. Sleaze has its consequence in the erosion of public trust, the growth of public cynicism, increasing political apathy, and deeply felt pessimism about the quality of public life and civic affairs. In turn, the low level of public confidence feeds the conviction that the people at large are not up to the job of ruling themselves, that inclusive democracy will lead inevitably to decline and decay.

If our thesis is correct, that an electronic republic opens the way for the public at large to become the actual fourth branch of government, then it is essential that urgent steps be taken to improve the quality of citizen deliberation in the public sphere. The last three chapters of this book suggest policy reforms in three critical areas: the individual and mass media, our political systems, and the quality of citizenship.

9

MEDIA REFORM— BACK TO THE FUTURE

Nothing could be clearer and simpler than the language of the First Amendment that protects free speech and a free press: "Congress shall make no law . . . abridging the freedom of speech, or of the press." Only fourteen words, and as Supreme Court Justice William O. Douglas noted, their meaning is plain: "Government shall keep its hands off the press."

Despite its unambiguous language, the history of First Amendment doctrine involving speech and the press is as complex and contradictory as the range of human discourse that the constitutional amendment was put in place to protect. The evolution of First Amendment law, said Harvard Law Professor Laurence Tribe, "is simpler to summarize than to comprehend." Trying to make sense of how the First Amendment should work in the future, especially during a time of radical change in communications technology, is daunting, but so is trying to understand the at-times mysterious and often contradictory ways of First Amendment doctrine during the past two hundred years.

Telecommunications technology alone has expanded to the point where, as we have seen, video today can be delivered in a variety of ways: videocassette recordings and discs, over the air television through broadcast stations, cable television, telephone line transmission, direct broadcast satellite (DBS), wireless cable (multichannel multipoint or local distribution service (MMDS or LMDS), and computers (CD-ROM). The future world of high resolution, interactive and digital transmissions will soon be here, permitting viewers "to order, access, store, and manipulate video, when and where they want it."

My own conviction is that the more complicated and diverse communications technology becomes, the simpler and more unambiguous our First Amendment protection should be. The electronic republic will best be served in the twenty-first century by returning to the late eighteenth century approach to the press that was specified in the Bill of Rights. Its content should be entirely free from "abridgement" by government. In that respect, tomorrow's telecommunications media should enjoy the same freedom as yesterday's print press. That freedom should hold no matter what form its content may take: whether print, sound, film, or tape; whether the message appears on television, computer, or movie screen, or is delivered via satellite, transmitter, microwave, cable, phone, fax, printing press, or soapbox.

Although written long before the advent of what we now know as mass media and certainly long before the arrival of personal telecommunications media, the First Amendment's centuries-old language, taken literally, should be the beacon for the future. The following principles should shape the nation's approach to free speech and a free press during the transformation to the electronic republic, no matter how the telecommunications environment may evolve:

- The First Amendment should apply to *all* media equally. The content of all media should be equally free from government intrusion.
- No *prior restraint* should be placed on any medium, no matter what its format.
- No restrictions should be imposed on *who* may publish or transmit information.
- The maximum possible diversity of media ownership and control should be sought.
- There must be universal access to the emerging transmission networks; a public sphere should be reserved for all citizens for civic information, discussion, debate, and decision making.
- There must be a free, independent, and properly financed system of public telecommunications: That system, I believe, should be supported at least in part by fees from commercial telecommunications service providers and spectrum auctions.

ALL MEDIA EQUALLY FREE

All media, whether print or electronic, should receive equal First Amendment protection; government should have no voice in regulating the editorial content of any medium. This is directly contrary to the practice of the recent past, during which the print press has operated largely free of government regulation while broadcast media have been both licensed and regulated.

More than a half century ago, Harvard Law professor Zechariah Chafee made the point that when new methods for spreading facts and ideas were introduced or greatly improved by modern inventions, writers and judges had not got into the habit of being solicitous about guarding their freedom. And so we have tolerated censorship of the mails, the importation of foreign books, the stage, the motion picture, and the radio. For centuries before that newspapers, books, pamphlets and large meetings were the only means of public discussion, so that the need for their protection has long been generally realized.

At their introduction, new media have at first consistently been licensed and regulated by government. It was true during the first century of the printing press and it was also true centuries later with radio and television, notwithstanding the First Amendment. Many reasons have been advanced to justify government regulation each time a new medium has come along. Some thought the new media initially too frivolous to be worthy of First Amendment protection—radio operators were looked upon as "the linear descendants of music halls and peep shows"; motion pictures were seen as nothing but "shows and spectacles," in business not to provide education or information but solely for profit. Others saw the new media as too powerful and pervasive to be left entirely uncontrolled and unregulated by government. Or regulation was thought to be required because spectrum space was scarce and only government could decide who could receive licenses to operate and who could not. Or the new media were believed to have the potential to endanger entrenched media interests, which, ironically, have not hesitated to demand that the government protect them from the threat of new competition, despite the First Amendment. Thus, newspaper publishers lobbied Congress and the FCC to prohibit radio from carrying news reports and were successful until the courts threw out that absurd restriction. Broadcasters lobbied to stop the spread of cable and succeeded for many years. And cable operators are doing all they can to get the government to prevent phone companies from carrying video programming.

All new media have been born under the cloud of some form of lesser First Amendment protection. Even though broadcasting, for example, long ago displaced the printed press as the major source of news and information for most Americans, it still does not enjoy the same degree of First Amendment protection as newspapers and magazines. Broadcasters have been required to cover controversial issues and to do so fairly (a requirement recently dropped), to offer those personally attacked or opposed in a political editorial the right of reply, to provide equal time to all candidates at the lowest rate, to afford access to purchased airtime to federal candidates, and generally to operate under a government-regulated, public interest, multiyear licensing scheme. Newspapers and magazines, meanwhile, are saddled with no such inhibitions; they are free to be as unfair, partisan, exclusionary, and frivolous as their owners and editors wish. The print press can ignore controversial issues, charge candidates any rate for advertising space, deny space to candidates altogether, be one-sided rather than even-handed, personally attack anyone, and give no one the opportunity to reply.

During the past century, the "press" evolved into the "media" and has taken many different forms, from hand-printed broadsheets to digitally transmitted telecasts to computer-generated on-line bulletin boards. Future technologies, as we have seen, will develop still more forms of communication. Yet until now the practice, wrote Justice Robert Jackson, has been to consider each new medium "a law unto itself," thereby justifying the "differences in the First Amendment standards applied to them." That means "a separate First Amendment test must be applied to each medium and a new standard developed with each technical innovation." Such an approach, said media scholar Ithiel de Sola Pool, "has led to a scholastic set of distinctions that no longer correspond to reality. As new technologies have acquired the functions of the press, they have not acquired the rights of the press." First Amendment decisions "reveal curious judicial blindness, as if the Constitution had to be reinvented with the birth of each new technology."

That doctrine should change. Whatever their form, the new media should enjoy the same full measure of First Amendment protection as the old-fashioned press. The government should keep its hands off content. Thus, no communications medium should be required to operate under standards and obligations imposed for decades on broadcasting by the government. The public interest, public trusteeship licensing approach interferes with editorial autonomy, has serious First

Amendment strains, and in any event has been a failure. In 1973, FCC Chairman Dean Burch told a broadcast industry group, "If I were to pose the question, what *are* the FCC's . . . controlling [public trustee-ship] guidelines, everyone in the room would be on an equal footing. You couldn't tell me. I couldn't tell you—and no one else at the commission could do any better." In 1976, Commissioner Glen Robinson described FCC regulation of broadcasting as a charade—a wrestling match full of fake grunts and groans but signifying nothing. Today, with stations getting automatic license renewals merely by sending a postcard to the FCC indicating they have complied with its requirements, the charade continues. Under broadcasting's public trustee licensing approach "the privilege to broadcast has been granted to friends of the government and withheld from its foes; efforts at censorship have been employed to back the political agenda of the party in power; and abuses have occurred with unfortunate frequency."

Further, the justification that channel scarcity requires the government to regulate the content of broadcasting no longer exists. Today there are 11,647 radio stations and 1,519 television stations; in all but a handful of cities, the local newspaper—not the local broadcaster—enjoys the monopoly position while the number of television outlets continues to expand. Currently, the average household has access to dozens of channels. By the end of the century, most people will be able to tune in to hundreds of channels or more. Or even more likely, no longer will there be channels at all; instead they will be replaced by a video dial tone of infinite capacity, enabling people to call up any programming, data, or telecommunications services whenever they wish. The time when three all-powerful broadcasting networks scheduled a narrow range of popular programming and held a whip hand over advertisers, affiliates, producers, and viewers has passed. Television scarcity, compared to print, no longer provides a rationale to regulate electronic media while letting newspapers and magazines alone.

In addition, the convergence of the media—both technologically and financially—is fast undermining our two-tier approach to government regulation. Newspapers like *The New York Times,* the *Wall Street Journal,* and *USA TODAY,* are transmitted nationally by satellite and also supply news electronically to on-line computer services and local cable systems. Their pages are often displayed on video screens. Increasingly, information and data are transmitted digitally and then converted into the medium of choice—onto wires, screens, or paper pages. Fax machines deliver print information using phone lines traditionally reserved for voice transmission. Books are being converted to compact discs and CD-ROMs. Phone companies, impatient to offer a video dial tone on their fiber-optic networks, aspire to provide television news, entertainment and sports, financial and other information services, and even classified advertising. Print, cable, broadcast, and satellite transmissions have become so integrated and overlapping that it is almost impossible to figure out why some segments of the press should be regulated differently from others, or, indeed, which segments belong in which category. In sum, the state should play no role in "abridging" or regulating the content of any communications medium, whether radio or telephone, newspapers or television, videocassettes or feature films, computer printouts or CD-ROMs.

There will be exceptions, as there are to any rule. But they should be held to the barest and most essential minimum. Undoubtedly, the government will continue to try to deal with the difficult and intractable problem of obscenity, but it is a mistake for government to be involved in issues of indecency or violence. For these concerns, as we shall suggest later, the media should be subject to independent criticism, peer pressure, and audience protests rather than government regulation. When it comes to restricting what children can view, rather than government intervention limiting what kinds of programs can be broadcast when children watch, the audience should be empowered to control their own children's viewing through lock boxes or other technical devices that restrict viewing choices at the set owner's option.

In sum, as the first principle, the goal for the future should be this: All media equally free.

NO PRIOR RESTRAINT

Second, there must be *no prior restraint* on what the media can produce and distribute. Only rare and limited exceptions to that rule should be tolerated, such as during a war or crisis when publication would create an imminent clear and present danger to life and limb, or in a courtroom when live transmission or instant publication might preclude a fair trial or impinge on a right to privacy. The presumption should be in favor of publication and transmission rather than suppression or prior restraint, no matter what the medium. Any criminal or civil violations or other damages—whether libel, slander, obscenity, plagiarism, or violations of privacy—that may be anticipated from publication or electronic transmission should be dealt with only *after* the offending matter appears, not before.

As Yale Law Professor Alexander M. Bickel said while defending the right of *The New York Times* to publish the Pentagon Papers, "A criminal statute chills, a prior restraint freezes." In the electronic republic, getting information and ideas in any form to citizens should take priority over whatever harm someone may think that information or those ideas will later cause, short of irreparable, life-threatening consequences.

NO LIMITS ON WHO MAY PUBLISH OR TRANSMIT INFORMATION

Third, government should impose no restrictions or prohibitions on *who* may publish, transmit, sell, or distribute information in any form through any medium. Anyone should have the right to communicate in any medium without prior approval by the state. Obviously, this would not preclude taking action against someone whose undue control or dominance becomes a bottleneck that prevents others from disseminating their messages. Nor will it assure everyone an equal public voice in the democratic process. The rich and powerful will always have more say and wider reach than the poor and weak. But everyone should have the right to spread the word (or picture) through any medium without fear of being silenced or excluded by government.

Again, there will be exceptions. Currently, for example, limits are imposed on foreign ownership of broadcasting stations—a restriction that is likely to continue although satellite technology has made it largely obsolete.

LOOKING BACK TO THE PAST

Those three classic First Amendment principles should undergird all communications policy for the future: No government role in content; no prior restraint; and no limits on who has the right to publish, transmit, talk, or distribute in any medium. How do these principles correspond to the First Amendment practices of the past?

In this century, the Supreme Court has treated the idea of free speech and a free press usually, but by no means always, with remarkable deference. On the Supreme Court, Justices Hugo Black and

William O. Douglas considered the First Amendment absolute and uncompromising; Benjamin Cardozo and William J. Brennan saw it as almost uncompromisable; and Louis Brandeis and Oliver Wendell Holmes thought the First Amendment superior to other, occasionally competing constitutional values.

Notwithstanding the First Amendment's plain language and exalted status, however, we have experienced a sharply divided, two-tier approach to the government's treatment of the press. As we have seen, the government's treatment has been different for different media. And it has also been different during times of war and crisis than in times of peace.

Throughout our history, vast disparities existed in the degree of press and speech freedom that was allowed during periods of perceived danger to national security. Justice Brennan made that point with considerable feeling and timeliness in a talk at the international symposium on free speech at Hebrew University in Jerusalem in December 1987. Brennan's keynote speech at the symposium happened to coincide with the outbreak of the *intifada*. As mentioned earlier, as president of NBC News, I, too, traveled to Israel to speak at that conference, although I actually spent most of my time out in the field, accompanying the NBC crews covering the Palestinian confrontations. In welcoming remarks to the guests at the conference, Israeli President Chaim Herzog and the Israeli attorney general used the occasion to blame the riots on the presence of television cameras, particularly our American network cameras. Young Palestinian radicals, they charged, were prompted to act up before the cameras in order to attract worldwide sympathy to their cause. If there had been no television coverage, the Israeli leaders insisted, the uprising would not have taken place. The Israeli attorney general, in particular, threatened to ban reporters and especially television crews from the scene if the riots continued. A free speech conference was a peculiar setting to make such threats, but on the other hand, the attorney general was talking to an audience that included the top-level news executives whose editorial decisions apparently had provoked his wrath.

Justice Brennan offered an extraordinarily eloquent and diplomatic response. He cited the questionable record of our own country in protecting civil liberties during times of crisis. "As adamant as my country has been about civil liberties during peacetime," Brennan said, "it has a long history of failing to preserve civil liberties when it perceived its national security threatened. . . . There is considerably less to be proud about, and a good deal to be embarrassed about, when one reflects on the shabby treatment that civil liberties have received in the United States during times of war and perceived threats to its national security." Brennan urged Israel to set a modern-day example for democracy throughout the world by behaving differently and preserving civil liberties even when under fire.

The American Supreme Court justice noted that the First Amendment had only just been adopted when the United States, thinking itself on the verge of war with France, enacted the Sedition Act of 1798. That act made it unlawful to "write, print, utter or publish . . . any false, scandalous and malicious writing . . . against the U.S. Government, Congress or the President with the intent to bring them . . . into contempt or disrepute." The Sedition Act was repealed several years later but never declared in violation of the First Amendment. During World War I, the Espionage Act of 1917 made it a federal crime to print or utter false statements made with the intent to interfere with the success of U.S. military forces or military recruiting in time of war. In 1918, the Espionage Act was amended to make it a crime also to "wilfully utter, print, write or publish any disloyal, profane, scurrilous or abusive language about" the U.S. form of government, its Constitution, flag or military forces and uniforms, "or any language intended to bring the[m] into contempt, scorn, contumely or disrepute." More than two thousand people were prosecuted for violations of this act, Justice Brennan said.

In 1919, even the peerless Oliver Wendell Holmes rendered all but meaningless his own famous "clear and present danger" test for speech during wartime. "When a nation is at war," Holmes declared, "many things that might be said in time of peace are such a hindrance to its effort that their utterance will not be endured so long as men fight." Holmes voted to uphold the conviction and prison term of a pamphleteer who had done nothing more than criticize the draft and question its constitutionality.

During the cold war in the mid-twentieth century, the Smith Act had made it a crime to distribute "any written or printed material advocating . . . the . . . propriety" of overthrow or destruction of the government with the intent to cause it to come about. As Justice Brennan concluded, while American public support for free speech and a free press has been widespread and widely celebrated, it also has been remarkably shallow, especially during times of crisis, when such protection is needed most. Each time a new national crisis has come to the fore, the First Amendment's free speech and free press clause has been diminished.

During my own years at both PBS and NBC News, I had seen much to confirm Brennan's concern about our nation's own practices during the cold war. I had experienced periodic efforts by high government officials to suppress stories because of their presumed dangerous consequences to some national or international crisis. When I came to PBS in 1976, the network was still suffering from the aftereffects of President Nixon's efforts to eliminate all of its public affairs programming on the grounds that it was too liberal. It took several years before we were able to reinstate a comprehensive schedule of public affairs programming on public television. Now public broadcasting is experiencing the same kinds of attacks from the right despite its predominantly middle-of-the-road and orthodox public affairs programs.

In December 1984, soon after I arrived at NBC News, Secretary of Defense Casper Weinberger called me out of an affiliate board meeting in Maui, Hawaii, in an effort to stop NBC from broadcasting a news report on Pentagon plans to launch a giant spy satellite. Speaking on an unsecured telephone line from Washington, D.C., Weinberger claimed that the national security would be jeopardized and the lives of secret agents in the Soviet Union could be lost if the story were revealed. (The secretary introduced himself to me on the telephone with the comment that he had been a journalist himself once many years ago, when he was editor of the *Harvard Crimson,* and realized how difficult it was to ask a newsman to kill an exclusive story.) In response to Weinberger's plea, I ordered our story postponed, only to see the same information published a few days later in the *Washington Post*. The launch was hardly a secret to the Russians, only to the American people. The Pentagon, it seemed, had decided to break the news of the spy satellite launch itself in a large-scale press conference and was unhappy that we were going to beat them to it. The secretary may have been misled by his own public information officers about the dangers of our news exclusive to national security and to our secret agents.

In May 1986, Director William Casey of the Central Intelligence Agency scared my elderly mother, who happened to tune in to a radio news report in which he threatened to prosecute me personally, as well as NBC News, for a story we had just run about an upcoming spy trial that, he complained, contained sensitive information. The information—that the Russians had succeeded in getting hold of secret data about our ability to tap into their leaders' private telephone lines—was already in Soviet hands, which was why the spy trial was being conducted in the first place. Casey's threat was simply a bureaucratic publicity ploy—an effort to distract the public's attention from what had been a rash of revelations about the CIA's failure to protect national secrets from being sold to the Russians. After Casey achieved the headlines that he wanted, blaming NBC News for the embarrassing problem, he dropped the matter, much to my mother's relief.

During the invasions of Grenada and Panama and the brief war in the Persian Gulf, the government placed unprecedented restrictions not only on all press reporting but also on all media access, including, in Grenada, an absolute ban on any coverage whatsoever. In every case, these clampdowns on free speech and free press were carried out with the enthusiastic support of the American people.

As we enter the twenty-first century, we have an opportunity to change all that, to set a course for the future, as Justice Brennan urged, by going back to the basic free speech and free press principles originally articulated in the First Amendment, even if not actually observed throughout our history.

In the interactive telecommunications age, as I have said, the three principles that need to be added to ensure that the first three underlying constitutional principles will work are: diversity of media ownership and control, some form of universal access to the public sphere, and a free and independent system of public telecommunications.

DIVERSITY OF MEDIA OWNERSHIP AND CONTROL

An electronic republic whose governance depends so heavily on an informed and knowledgeable public requires an open market for opinions and ideas. The best way to achieve that is to ensure the widest possible diversity of media ownership and, in the new telecommunications realm, the widest possible access by information providers.

The problem today, as we have pointed out, is that media ownership is going through what appears to be a continuing and accelerating process of consolidation and convergence. The examples are legion. Capital Cities owns ABC television and radio stations and networks, major newspapers and magazines, and an increasing number of the most popular cable program services. Time Warner owns major national magazines, is the second biggest operator of cable television systems, holds all or part of the most important cable program networks including HBO and CNN, owns a leading movie and television production company, owns several of the largest book publishing companies, and the biggest recording companies. Rupert Murdoch's News Corp. owns television stations, a broadcast network, book publishing companies, magazines including *TV Guide,* cable channels, a major movie studio, and newspapers and satellite channels not only in the United States but throughout the world. Even as the number of television and radio channels has proliferated, ownership of our mainstream media, both print and electronic, has shrunk drastically. Fewer than two dozen large, politically powerful companies control delivery of most of the news and information we receive.

According to one analysis, fourteen dominant companies own half or more of the daily newspaper business, three companies own more than half of the magazine business, three in television, six in book publishing, four in motion pictures. As a result, ironically, as more media outlets have opened up, greater financial barriers to entry have been raised to new and independent sources of programming and information. If the present rate of shrinkage continues, the prediction is that by the next decade a mere half-dozen companies will own the largest newspapers, magazines, television and cable outlets, book publishers, satellite services, and movie studios that reach the majority of the nation's audiences and capture most of the nation's media revenues. And that consolidation is rapidly going global.

In the U.S. cable industry, fewer than ten cable operators now own the systems that reach the majority of the nation's cable subscribers. The same companies also own controlling interests in the most important and most popular cable program services and production companies whose products appear on their cable systems. In order to gain carriage on cable systems, new program services feel obliged to offer shares in their business to the most important cable operators. Similarly, as we have seen, cable companies have been known to exclude program services that compete with those they own. With TCI and Time Warner, the two biggest cable system owners, both holding major financial interests in CNN, no other cable news service can gain entry into their systems.

Technology has made it possible for many voices and many interests to be served. But vertical and horizontal concentration and consolidation continue to reduce the number of major media owners. In addition, the combination of new communications technologies and the accelerating pattern of mergers and buyouts of companies with existing technologies have put exceptional power into the hands of international financiers and money managers, who have extraordinary influence on the development of the media companies themselves. The problem of a few corporate media owners holding the power to decide what the entire nation can see, hear, and read did not exist in 1791 when the First Amendment was added to the Constitution to protect the individual soapbox orator, broadsheet printer, and pamphleteer from the oppressive hand of government.

Today, the mainstream media companies depend heavily on government for their ability to own multi-million dollar broadcast licenses and cable franchises, and utilize exclusive satellite paths. And in return, legislators depend heavily on the mainstream media for their large-scale financial contributions and favorable press coverage. With legislative and regulatory decisions having life-and-death financial consequences to telecommunications companies by determining, for example, who can own which communications paths, the media rank among the nation's heaviest political contributors. Seats on the House and Senate Communications subcommittees are highly prized precisely because they consistently generate large political contributions from the nation's major media interests. Corporations involved in the escalating race to acquire media properties seek not only expanded profitability but also increasing influence.

Far from being independent and wary of one another, big government and mainstream media work hand in glove with each other. These days, the free press clause of the First Amendment is invoked more often to protect the property rights and special financial interests of media owners from antitrust and other regulatory restrictions than to protect full and free public debate or to defend the expression of unpopular opinions. Commenting on the degree to which the First Amendment is being used as a shield for media business interests, Cardozo Law School Dean Monroe Price complained that "the victory of Tom Paine is being corporatized. The soap box is being replaced by the mall."

The growing concentration of media ownership, leaving only a few corporate gatekeepers in charge of most mainstream outlets, has led some to worry more about corporate power over free speech than about the government's abuse of it. Paradoxically, in the name of First Amendment values and goals these critics call for more government regulation of the media not less in order to ensure fairness and balance in the media and a level playing field in public debate. They argue that if free and open debate on issues of public importance is necessary for the healthy functioning of a democracy, then as the power to shape opinion, or decide which opinions are heard, is concentrated in fewer hands, democracy will grow less healthy. My own conviction, however, is that any regulations to limit concentration of media ownership should focus not on content—as with the traditional requirement of broadcasters to program in the public interest—but on antitrust enforcement and, where appropriate, on open access requirements to limit the power of the media gatekeepers to restrict the flow of information.

We have become far too tolerant of media concentration and gigantism. It may well be that some of the problem will solve itself. Top-heavy companies do tend to fall apart of their own weight, as happened with the proposed sale of TCI to Bell Atlantic and with the unwieldy media conglomerate assembled by the late Robert Maxwell. And new technologies have a way of wiping out the dominance of the old, as happened when the rise of television caused landmark changes in radio and the rise of cable caused major shifts in network television. Where the problem of media concentration continues, however, as it seems to be doing throughout the industry, the most effective way for government to deal with the issue is through vigorous antitrust enforcement. Antitrust laws should be enforced against media oligopolies on both the transmission and content sides of the communications industry, with the goal of achieving maximum diversity of media ownership. The top telecommunications operators should be limited in the amount of programming they themselves can own, produce, and decide to distribute through their own facilities. The three biggest cable system operators should not be permitted to own so many of the dominant cable program services they choose for their own cable systems. Newspaper owners should not also own television stations in the communities in which they publish. A television broadcaster or the sole newspaper in a community should not be allowed to program the cable company's local news channel in the same community. It is vital that people get information on local issues from diverse sources. A single corporation should not be able to control all or even most of the mainstream media in any one market. GE should not be allowed to own both NBC and the only national financial news service on cable, as well as major cable program series. Nor should the First Amendment stand in the way of taking steps to prevent a handful of people or companies from wielding too much power through their concentration of control over the mainstream media.

Two crosscurrents are at work in the new media environment. One is technological, empowering individuals to receive and distribute the information they want when they want it. The other is economic, reducing the number of multimedia owners of the mainstream means of communication. Bottom line, it is vital that all information providers have access to electronic transmission the way all print publishers have access to the mails, fax, and highways to distribute their wares. When telephone companies have the capacity to transmit video, they should provide video distribution to all who seek it on a first-come, first-served basis (which is what common carrier status requires). Similarly, cable operators should be required to lease at least 10 to 15 percent of their capacity to independently owned commercial providers, using compulsory arbitration to decide who gets access if there is no agreement on terms. In the electronic republic, there needs to be assurance of a media environment of wide-open access to send and receive. As we remarked in Chapter 4, quoting Justice Potter Stewart, the press is "the only organized private business that is given explicit constitutional protection, which also gives it special privileges *and* responsibilities." When it comes to property rights, however, the First Amendment gives the press no special privileges to create an information bottleneck or to be exempt from reasonable antitrust provisions against oligopolistic or near-monopoly control. The only special privilege it gives the press, and should give all media, whether print or electronic, is the invaluable right to say whatever it pleases.

We must distinguish, in short, between the *medium,* on which reasonable limits of ownership can and must be imposed, and the *message,* which must be kept entirely free from government interference, regulation, and restriction.

That is easier said than done. To be realistic, government regulations that impose limits on media property ownership inevitably affect media content. Indeed, gaining more diversity of content is the very reason for seeking greater diversity of media ownership. However, rules affecting media ownership have been capricious and continually changing. Newspaper companies can own as many daily papers as they wish but are restricted in the number of broadcast properties they can control. Cable operators want no limits placed on the number of cable systems or cable program services they can own and also plan to offer telephone services through their cable lines. But the cable companies have fought hard to ban telephone companies from providing any video programming through phone lines. Now broadcast television networks can own a high proportion of the programs they carry and can even own cable systems. And the way is being opened to ease controls on telephone companies, allowing them to launch a vast array of video entertainment and information services. The more the merrier, as long as the media owners do not have the power to short-circuit media entry by nonowners simply to protect their own expanding interests from added competition.

As one now-retired broadcaster, a former chairman of the NBC board of affiliates, put it, "The FCC, the SEC and the anti-trust division of the government must work to reduce monopoly control of communications—not to block another William Randolph Hearst and his dictatorial policies, but to insure a system that should prevent what is in the public interest from becoming a cash box. Bottom line thinking simply precludes public service investment. Television licenses do have great value and they should carry responsibilities. Currently, they enjoy the financial values but, for the most part, the organizations owning the licenses ignore the responsibilities. Without public discipline, they will continue to do so."

The basic principle of diversity of media ownership and freedom of entry should serve as an essential guideline for national media policy. The American people should receive information from as many different sources as possible.

UNIVERSAL CITIZEN ACCESS

In order to exercise their rights and obligations as citizens in an electronic republic, every citizen will need access to a basic set of interactive telecommunications tools, just as today everyone who votes needs access to a ballot. In the future, national policy should ensure that all citizens will have available to them the basic telecommunications equipment they will need to participate in the public sphere. Tomorrow, everyone should be able to *receive* electronically essential civic, educational, health care, and similar information. And everyone should also be able to *transmit* electronically his or her votes, views, and ideas to elected officials as well as to other citizens.

That means a national policy granting universal access by all citizens to the evolving broadband digital transmission networks. Subsidies may be required to provide essential public telecommunications services for the poor, as has been done in the past with rural electrification and universal basic telephone service. In January 1995, House Speaker Newt Gingrich suggested in testimony before the Ways and Means Committee that perhaps the poorest Americans should be given a tax credit to buy a lap top. "I'll tell you, any signal we can send to the poorest Americans that says, 'We're going into a twenty-first century, third-wave information age, and so are you, and we want to carry you with us,' begins to change the game," Gingrich said. Later, Gingrich admitted it was a "stupid" idea (if only because tax credits are useless to the poorest Americans). But the principle is right for the electronic republic, even though it is still too early to know exactly what basic telecommunications equipment every citizen will need in the future.

In the past, of course, no public subsidy was needed for the poor to purchase television sets. Heretofore, television was regarded, accurately, as an appliance used primarily for entertainment. The result is that today, without subsidy, more of the nation's households have access to television than have telephones or indoor plumbing. It will not be long, however, before today's television sets evolve into tomorrow's telecomputers—advanced interactive digital receivers. Their price should fall dramatically, in line with Moore's Law that the number of transistors on a chip doubles every eighteen months and the price declines proportionately. If the poor will require help in gaining access to tomorrow's telecomputers in order to exercise their rights of citizenship in the electronic republic, a subsidy should be designated for that explicit and essential public purpose and targeted to those in need.

HOW TO PAY FOR PUBLIC TELECOMMUNICATIONS SERVICES

A nonprofit Public Telecommunications Trust Fund should be established to help finance not only barebones universal telecommunications service for the poor if necessary but also the essential content and substance of public service telecommunications—the critically needed quality programming, civic information, educational materials, and data that should flow through the public sphere.

The State of Maryland's funding of universal phone service offers one model of how basic public telecommunications equipment for the "have nots" might be paid for. The state simply decides each year on the amount of the subsidy the poor who live in Maryland will receive for each family's basic telephone service. The state then authorizes those on welfare to go to the telephone company to get that service. In turn, the telephone company, which is subject to the usual gross receipts tax, deducts the amount of the subsidies when it remits its tax to the state.

France set another kind of example early in the decade by distributing its "minitel" system to every French household that wanted it. Connecting the telephone line to the television screen and computers, minitel enables French citizens to look up phone numbers and transportation schedules, order train and plane tickets, and carry out a great variety of other information services, for a fee. In principle, the minitel interactive device can be designed to be used for political purposes as well, enabling citizens to express their views, respond to surveys, cast electronic ballots, and exchange information and data.

My own conviction is that in the electronic republic, the cost of providing public telecommunications services, including equipment for the poor as well as content for all citizens, should come from several sources—annual spectrum use and cable franchise fees, a modest tax on the sales price of station licenses and cable franchises, revenues from auctions of unused spectrum space, voluntary public contributions, corporate and foundation underwriting, and even limited or "clustered" advertising aimed at adult viewers of nonprofit public telecommunications outlets during certain specified hours. However, the cost of public service telecommunications should be borne primarily by commercial telecommunications providers such as broadcasters and cable operators who, typically, reap large profits from using public rights-of-way. Currently, public broadcasting is financed in part by congressional appropriations from general tax revenues. However, government-mandated money for a Public Telecommunications Trust Fund, which would supersede the current public broadcasting federal appropriations, should come not from general federal tax revenues but from fees paid by commercial telecomunications providers. The cost will be passed on to their customers, who are all of us, in any event. Notwithstanding the current trend to cut back rather than increase financial support for public broadcasting, if democracy is to thrive in the electronic republic, the tools for citizen participation and the information needed for informed public judgment must be made available to all. The public sphere cannot be left entirely to the private marketplace.

The time is long overdue for commercial broadcasters and other commercial spectrum users to pay a modest price for their use of public spectrum for their own private gain. "Congress could reasonably establish such a fee based on a percentage of gross revenues, say, 1 percent for radio and 2 or 3 percent for television." The sums obtained—$600 to $800 million a year—augmented by a portion of the revenue from auctions of unused spectrum and by a portion of the fees paid for cable franchises, would finally help resolve the perpetual funding problems of public telecommunications.

Cable operators already pay fees to local governments for their franchises, but the franchise fees are absorbed into the municipalities' general funds rather than set aside for public service communications services on cable. Congress should mandate that 1 or 2 percent of the cable franchise fee be dedicated to the Public Telecommunications Trust Fund to help support programming for the access channels on cable for public, education, and government use (the so-called PEG channels). The report of the House of Representatives on the 1984 Cable Act described these channels as "often the video equivalent of the speaker's soap box or the electronic parallel to the printed leaflet. They provide groups and individuals who generally have not had access to the electronic media with the opportunity to become sources of information in the electronic marketplace of ideas. PEG channels also contribute to an informed citizenry by bringing local schools into the home, and by showing the public local government at work." The problem is, however, that the PEG channels are found in only about 15 percent of the nation's cable operations and they have been left largely unfunded. Congress should remedy both of those deficiencies.

Broadcast licenses, cable franchises, direct satellite broadcasting frequencies, and cellular phone frequencies have consistently ranked among the most lucrative government giveaways in the nation's history. High television station profits—a 50 percent profit margin has not been uncommon—are derived largely from the simple fact that commercial broadcasters receive exclusive licenses to operate on the public airwaves. When broadcasters sell their television stations for hundreds of millions of dollars each, most of that sales value lies not in the stations' transmitters, studios, office furniture, business goodwill, or program inventory, but in their commercial broadcasting licenses themselves. Similarly, in the cable business, three-fifths of the average cable system's equity—with values currently averaging more than two thousand dollars per subscriber—lies in the value of the government-granted cable franchise itself.

Broadcast station licenses and monopoly cable franchises cost their original owners nothing except legal fees. The exploitation of those licenses and franchises brings in billions of dollars in revenues. Broadcasters got their licenses originally by committing to serve the public interest first and put profits second, a commitment the great majority of them have totally failed to keep. On top of that, with deregulation, the broadcasters' original requirement to serve their communities' needs and interest, in exchange for receiving their windfall licenses to operate radio and television stations, barely exists any longer.

True, television stations still have certain minimal legal obligations to schedule children's and community-issue oriented programming, which most of them barely fulfill or ignore altogether. The requirements of the Children's Television Act of 1990, to carry at least some programming designed to meet children's "educational and informational needs," are honored more in the breach than the observance. In their reports to the FCC, many TV stations list blatantly commercial entertainment shows, such as *The Jetsons, G.I. Joe, Yo Yogi!,* and *Leave It to Beaver,* as "educational and informational" in an attempt to satisfy the law's requirements. One station actually claimed to the FCC that by capturing a "bank robbing cockroach," the hero of *Yo Yogi!* educated children to "use . . . [their] head rather than . . . [their] muscles."

Broadcasters and cable operators will argue that while all cable operators and station owners originally were awarded their licenses and franchises free, many subsequently bought theirs in private sales for substantial prices. But the point is that all commercial telecommunications providers that profit from using public rights-of-way should pay a modest public dividend for that privilege. And in the case of broadcasters, a license fee should also be considered a small repayment to the public sphere for being relieved of their long-standing public service responsibilities, thanks to deregulation. As former FCC General Counsel Henry Geller pointed out, "Commercial broadcasters should welcome and support this legislative opportunity [for spectrum fees], which would be highly beneficial to them. They would have long-term franchises free of public service scrutiny, the renewal process, including petitions to deny and comparative renewals, all fairness controversies, and anti-trafficking regulation. Because the public does not generally distinguish between cable and over the air television, the regulatory scheme for both should be similar: both would then have long-term franchises and franchise fees (although the broadcast fee may be lower than the cable fee), and neither would be scrutinized for public service content."

FCC Chairman Reed Hundt said in an interview with *Broadcasting and Cable* that "the time has come to reexamine, redefine, restate, and renew the social compact between the public and the broadcasting industry." The Chairman also commented that if the compact is broken (which it surely is), the broadcasters will face a spectrum usage fee. The time has come to put such a fee in place.

Similarly, after they become profitable, an appropriate annual telecommunications provider fee should be charged for video telephony, direct broadcasting via satellite, and other new distribution systems that use the public spectrum. In addition, whenever commercial licenses and franchises are sold, a small telecommunications sales tax should be imposed, to go to support the Public Telecommunications Trust Fund. A bill containing such a provision was originally introduced in the Senate by South Carolina's Ernest F. (Fritz) Hollings, but was quickly bottled up in committee and killed after considerable pressure from the National Association of Broadcasters.

Ample precedents exist for imposing spectrum use fees and cable franchise taxes for public telecommunications purposes. Ranchers, lumber companies, and offshore oil drillers pay fees to the government for their use of public lands and waters. These fees, in turn, help pay the costs of the public's use of these resources. Truckers are taxed to pay for the upkeep of the roads and highways they use. In some areas, developers who build on public property are assessed special fees to help pay for historic preservation. So it should be for the commercial exploitation of the public spectrum.

Absent such a public service commitment and the necessary financing to support it, even the increasing abundance of new electronic media in the telecommunications age will not, by itself, sustain an informed citizenry and a free and open deliberative democratic process. Thomas Jefferson remarked, in a little known postscript to his famous comment that he'd prefer newspapers without government to a government without newspapers, "But I mean that every man should receive those papers and be capable of reading them." To accomplish that goal in modern times we need to establish a basic interactive public telecommunications service that will be available to all. To paraphrase Jefferson, that means every citizen should receive those services and be capable of using them.

A FREE AND INDEPENDENT PUBLIC TELECOMMUNICATIONS SYSTEM

To make available to all citizens of the electronic republic the information they will need for sound deliberation and reasoned civic judgments, it will be essential to strengthen and, indeed, reinvent the nation's public broadcasting system. Noncommercial public radio and television should be converted into a vigorous, independent, well-supported local and national interactive public telecommunications system—democracy's great electronic forum.

Currently, America's public broadcasting stations, while weak and always starved for funds, are capable of reaching virtually every member of the radio and television audience in almost every community in the nation. With its unique ability to broadcast not only national but also local and regional programs, public broadcasting is ideally positioned to serve as the electronic republic's town square. In 1994, not long after becoming president of PBS, Ervin Duggan proposed that public television initiate a major "Democracy Project" that would strive to do just that, within the limits of PBS's declining budgets.

That proposal is in line with the aspirations for public television expressed by the First Carnegie Commission back in 1967: "Public Television programming can deepen a sense of community in local life. It should show us our community as it really is. It should be a forum for debate and controversy. It should bring into the home meetings . . . where major public decisions are hammered out, occasions where people of the community express their hopes, their protests, their enthusiasms, and their will. It should provide a voice for groups in the community that may otherwise be unheard."

For the new century, however, American public broadcasting needs to be entirely rebuilt and rejuvenated. It should be integrated with cable's local public access channels and with satellite-delivered public affairs channels, interactive computer bulletin board services, multimedia information data bases, and the rich variety of on-line new telecommunications delivery systems now in development. Locally and nationally, a multifaceted new public service telecommunications system could provide a nourishing, vital mix of electronic town meetings, political debates, issues forums, public affairs documentaries, public dialogues, interviews with government leaders and officials, press conferences, and public participation programs, as well as reports and analyses by political scientists, historians, economists, and experts of every political hue, representing every point of view.

Ample free time on the air would be opened up for political parties, labor unions, business advocates, civic associations, community action groups, public interest groups, and citizens' organizations of all kinds. Candidates for major local as well as national offices would have powerful outlets to make their case to constituents without being forced to pay exorbitant charges for television and radio time.

Cable's C-SPAN offers an inspired example of how even a modest, fully dedicated public affairs network, aimed at a limited audience, can still elevate the level of political dialogue. The brainchild of Brian Lamb, a former member of the Nixon White House staff, C-SPAN began in 1979 by televising verbatim the daily sessions of the House of Representatives. From the beginning, production costs of televising the sessions have been borne by the House, and the video cameras in the House chamber have been operated under the control of congressional employees. When the Senate eventually followed the House's example and voted to allow cameras into its chamber as well, C-SPAN expanded into a second cable channel, operating under the same rules.

I was peripherally involved in a somewhat obstructionist way in the start-up of C-SPAN. Lamb had come to PBS in 1978 to seek our help in transmitting the House sessions on the new public television satellite and to explore the possibility of even running some congressional segments on PBS itself. (PBS had led the way in building America's first broadcasting satellite distribution system in the late 1970s.) I rejected the idea for PBS on the grounds that C-SPAN's congressional programming was paid for directly by Congress and produced under the control of congressional staff members and, ultimately, of the members of Congress. I feared that the journalistic and programming independence of PBS, which was still merely a fledgling network, would be compromised if we got involved directly with C-SPAN, much as I admired its mission.

It was the right decision for PBS to turn down the C-SPAN approach, and Lamb, a dedicated and determined soul, made the C-SPAN idea work without PBS's help. Today, while Congress still pays for and oversees the televising of its own deliberations, C-SPAN also receives monthly fees from cable operators who recognize its importance to the nation and its political value to the cable industry. Under Lamb's modest but effective direction, C-SPAN has broadened its reach, supplementing gavel-to-gavel coverage of Congress with coverage of committee hearings, issues seminars, political forums, intelligent interviews, governors' state of the state messages, candidate speeches in full, and the like. Enabling cable viewers to see government unabridged, C-SPAN has become an invaluable service to democracy. While rarely offering particularly exciting or sensational fare, C-SPAN can still be engrossing to anyone with an interest in politics, although its appeal extends far beyond—to those with a special interest in whatever issue is being debated, to channel surfers, and increasingly to the curious public at large. Today, C-SPAN's cable service is seen in the homes of more than half the nation. While its audiences are small by broadcast television standards, it is still watched by almost 25 percent of all cable subscribers at some point each week. C-SPAN's impact comes not primarily from the number of viewers it attracts but from the nature of those viewers—politically motivated and interested citizens with a higher than average voting record who are not shy about communicating with their elected representatives. C-SPAN offers an important working model that helps fulfill television's potential to deal with politics in a responsible, serious, high-minded way.

Public television and radio can reach the entire nation and not just the 60 percent of the nation with access to cable. Their local *and* national broadcasting schedules offer an even more democratic vehicle for expanding the audience's access to substantive civic information, deliberation, and debate. Their programs certainly do not achieve massive audiences comparable to those the commercial networks routinely attract. But PBS's *MacNeil/Lehrer Newshour* and *Washington Week in Review* and NPR's *Morning Edition* and *All Things Considered* demonstrate that public broadcasting has the capacity to achieve meaningful influence and impact nationally with fair-minded, responsible programming.

Public broadcasting should make a significantly greater commitment than it does now to carrying critically important public affairs programming in prime time, supplemented by major public affairs efforts on public cable and satellite channels and on other new telecommunications formats and vehicles. In the words of the Twentieth Century Fund Task Force on Public Television, issued in 1993, "the broad national values, the linkages among educational experiences, the in-depth coverage of public issues, and the common cultural experience that the best of public television can offer seem of greater value than ever."

A working prototype of what is already possible, using the new interactive telecommunications media, is the proposed Democracy Network, a new digital system of interactive, multimedia political communications being developed at the Center for Governmental Studies in California. The Democracy Network would use cable, telephone, television, and computer transmissions. It could have a profound effect on election campaigning, on campaign financing, and on citizens' ability to gain information about civic affairs.

The Democracy Network proposes to allow voters, through their TV sets or computers:

- to watch video, audio, and textual materials on demand from political candidates and ballot measure groups by simply clicking a TV remote control or using a voice command,
- to view TV, radio, and newspaper materials and editorials on candidates and issues,
- to talk with candidates and voters in an electronic town hall,
- to receive multilingual materials for non-English-speaking audiences.

Once installed in cable and video systems, the network hopes to "increase voter participation, especially by poor, young and new voters; decrease campaign costs and financial disparities between candidates; encourage candidates to devote more attention to substantive issues; and demonstrate to elected officials the value of incorporating free voter information into the new definition of 'universal service.' "

Similarly, on the local level, in the late 1980s, the City of Santa Monica established its own public access computer on-line network that allows the one-third of Santa Monica residents who have computers in homes or offices, or who can use about 30 terminals in public places, to:

- access databases about city government and schools, including lunch menus, park schedules, and events listings.
- read the city's public library catalogue and borrow books by mail.
- send and receive electronic mail from city government. (Senior executives in each department are required by the city manager to respond to each message within twenty-four hours.)
- conduct transactions with the city, such as getting a building permit, checking on licenses, and learning deadlines.
- engage in conversation with anyone else on the network. (For example, a conversation among citizens on the Santa Monica network led to the establishment of a group that provides showers and laundry facilities to homeless persons.)

Will such programming ever attract large-scale audiences? Hardly ever, except perhaps when focusing on especially contentious and critical issues. But cumulatively, a vigorous nationwide multimedia public telecommunications system with local outlets in virtually every community serving as democracy's electronic forum, could contribute enormously to improving the quality and seriousness of the political debate. In the electronic republic, public service telecommunications should become the centerpiece for responsible civic deliberation—the electronic public sphere—the equivalent of the ancient gathering place for the citizens of Athens on the hill called Pnyx.

The only other option, to rely entirely on the commercial marketplace to decide how much and what kinds of news and information we receive, will offer a far from satisfactory alternative. As we have seen, the marketplace itself severely limits the quality of information and diversity of ideas. It looks upon people as consumers, not citizens. It favors audiences with money to spend. The marketplace caters to the requirements of advertisers. It offers what has instant, immediate, and arresting appeal, what is presold, easy to take, and familiar. It emphasizes the conventional and predictable at the expense of the experimental and controversial. The marketplace, for all its responsiveness to public demand, often ill serves the poor and minorities. It looks upon children only as customers. It has little incentive to fulfill the public's need for civic education and serious information about public affairs and controversial issues.

Informational and educational programs for children and minorities and about serious public affairs, culture, and the arts have always required some form of public subsidy and private patronage. In that, they are no different from our libraries, museums, performing arts companies, schools, and universities. This will continue to be true no matter how many channels will be added to our television sets or how many competing media will come into existence. Public television's *MacNeil/Lehrer Newshour* and *Firing Line* with William F. Buckley, public radio's *All Things Considered* and *Morning Edition,* and cable's C-SPAN could not survive in a purely commercial broadcasting environment, even with channel abundance. Indeed, channel abundance, by fragmenting audiences, increasing competition for advertisers and viewers, and shrinking the public sphere, is more likely to diminish rather than improve the quality of programs and dilute rather than strengthen in-depth journalism.

To insulate a strengthened public service telecommunications system from political interference, the Public Telecommunications Trust Fund should be a nonprofit, nongovernmental enterprise, modeled on the current Corporation for Public Broadcasting. CPB was established to insulate public radio and television from political pressure. It currently receives the federal appropriations for public broadcasting and is prohibited by law from producing programs or operating a network itself. CPB's board of directors, which was intended to consist of distinguished Americans but has too often been populated by political lightweights, is nominated by the president and confirmed by the Senate. The new Public Telecommunications Trust Fund would be better served if its board members were nominated by the president from a group of candidates recommended by a blue-ribbon panel of university presidents, leading writers, artists, scientists, and citizens of accomplishment.

By law, most of the money CPB now receives from the federal government is paid out in the form of revenue-sharing grants that go directly to local public television and radio stations. This has the advantage of shielding the funds from Washington's direct political influence. But it has a liability as well, that too little federal money gets spent on the product—the production of programs—and too much gets spent on overhead to support duplicative, overlapping local stations. The current system would be vastly improved if, by law, most of the federal funds devoted to public service telecommunications were earmarked for the product—programming, information, and data services and software content. Local stations and telecommunications facilities should be supported largely out of local contributions and appropriations from local communities. This was one of the major recommendations of the Twentieth Century Fund's Task Force on Public Television, on which I served and whose findings I fully endorsed.

Additional financing for the new public service telecommunications system should come from public-spirited private sources in the form of voluntary contributions from members of the public, corporate underwriters, and foundations, as it does today. Under the proposed new system of dedicated public financing from spectrum fees and auction sales for public telecommunications, corporate funding should no longer play the dominant role in programming, as is now the case. In my view, it is essential for the public telecommunications system of the future to maintain a healthy balance between private and governmental financial support. Two elements are critical in designing a new and expanded public telecommunications system, suitable for meeting the needs of the electronic republic: First, government should play absolutely no role in determining its content. Second, sufficient funding should be provided to do the job properly.

That the system must be insulated from political interference is easy to say and difficult, but not impossible, to accomplish. The influence of politics on programming can taint the government's financial support for public telecommunications just as it can taint government support for education, libraries, museums, and the arts. Public broadcasting has had its own share of problems in this regard. The virulent controversy over "Death of a Princess" cited earlier was only one such instance. But continuing vigilance and a system designed to minimize political influence can effectively limit government to its proper role in serving First Amendment values while at the same time providing essential information to all the nation's citizens. The major problem of public broadcasting so far has not been too much government meddling but too little financial support.

Diversity of funding sources from both public and private contributors will help insure the independence and integrity of the Public Telecommunications Trust Fund. In addition, the comparatively insulated sources of federal funding proposed here—franchise and spectrum use fees, spectrum auction fees, and revenue from station and franchise property sales—will help insulate the new public telecommunications system from the political pressures that are inherent in the annual congressional appropriations process.

Can a well-funded public telecommunications system be reconciled with the First Amendment principle of a press free of government interference and involvement? The answer is yes. Democracy is the primary value that was meant to be served by the First Amendment, and effective democracy "is a value that would be difficult [if not impossible] to attain without an educated [and informed] electorate." As communications lawyer and theorist Robert Corn-Revere noted, this means that "there is an important role for public policy. The creation and support of schools, universities and libraries are affirmative acts that are appropriate for government to take that directly serve this First Amendment value. This legitimate role extends as well to electronic communications." The government should fulfill that role by creating a viable Public Telecommunications Trust Fund for the electronic republic in the new information age, as it did in the 1970s when it created an American public broadcasting system, weak, late, and inadequate though it has been.

10

NONGOVERNMENTAL AND OTHER REFORMS

The First Amendment prohibits or at best severely inhibits what government can do to improve the quality of privately owned commercial media. But there is nothing to stop *non*governmental pressures from being brought to bear to improve the nation's dominant providers of news, information, entertainment, and communications. Since American tradition, fortunately, largely rejects government intrusion in this area, other steps will have to be taken to raise the mainstream media's intellectual and cultural standards and sense of civic responsibility.

Surveys indicate the American people are growing increasingly dissatisfied with the performance of the news media, especially television. Viewers also consider newsmen and newswomen, like politicians, to be members of the insiders' group, the "special interests"— part of the problem, not the solution—"them, rather than us." Still, except for occasional threats to boycott elements of the press or their advertisers by archconservatives convinced that the press is hopelessly liberal and prone to favor Democrats, the general public has been remarkably tolerant of and complaisant about the mainstream news media.

It is time for concerned citizens—readers, listeners, and viewers— acting individually and in organized groups, to push and prod commercial broadcasters, cable and satellite operators, video producers and distributors, and print publishers of all kinds to vastly improve the quality of their performance. In a free market economy, people get the media they deserve. If they are dissatisfied with the media's performance and do nothing to change it they have nobody to blame but themselves. Public interest groups, good government associations, and civic organizations should lead the charge by insisting that the commercial media do far more than they do now to enrich culture, expand intellectual resources, and, especially, produce responsible, informative, and useful news and public affairs programs.

As one experienced broadcaster observed, "The advertising agency and the ratings drive programming. The best alternative to gain the immediate attention of a local station management is to write a letter to, or file a complaint with, the FCC. While no one with any sense assumes the FCC is worth a damn, the fact remains the bureaucrats must ask questions, and questions must be answered, and generally answered by lawyers who cost a bundle of money! Now, the problem of a station manager is money—the bottom line. The FCC is a paper tiger of almost useless value, but it can cost a lot of dough to a station operation." Advertiser and audience boycotts, vigorous letter-writing campaigns, picket lines, stockholder complaints, and other legal and appropriate instruments of effective public protest should be mobilized in a full-scale effort to improve dramatically the level of corporate media citizenship.

In the mid-1980s, NBC News began to videotape the most interesting and thoughtful viewer criticisms and reactions to our news broadcasts, to be carried as regular segments on *Today* and *Nightly News*. To my surprise, however, the anchors and producers of those programs objected so vehemently to airing the critical audience segments that the project had to be dropped. Too many journalists have become celebrities with glass jaws; good at dishing out criticism but notoriously bad at taking it on the chin themselves.

A few dedicated citizens can have astonishing impact on the media's performance. Peggy Charren, the energetic founder of Action for Children's Television (ACT), operating out of Cambridge, Massachusetts, almost singlehandedly has pushed quality children's programming to the top of the nation's priority list and brought about enactment of the 1990 Children's Television Act. Not that the quality of children's programming today is at all satisfactory or even tolerable. But without Peggy Charren, children's television would be even worse than it is now.

At long last, the academic world has begun to take seriously both the effects and the possibilities of electronic communication, especially television's influence in the public realm. One colleague told me that he had once complained to a group of educators about how they first refused to recognize television and then later simply demeaned it. One educator, he said, "suggested I take a turn at medieval history and see with what contempt they greeted movable type."

Traditionally preoccupied with high culture, disdainful of popular entertainment and the mass media, and accustomed to dealing primarily with print, the nation's academics once studiously ignored what was happening to our democracy and our society through electronic communications. Today, that appears to be changing. The proliferation of media studies centers at leading universities can play an increasingly important role in fostering critical self-examination of the media. Columbia, Harvard, Miami, UCLA, Southern California, Syracuse, Northwestern, Vanderbilt, Washington, Washington State, Pennsylvania, and a growing number of other universities now house academic centers entirely devoted to examining the media's role in contemporary society. Most have been funded by media interests. Even so, these scholarly academic centers should do far more than they currently do to offer sustained, knowledgeable, tough-minded academic criticism and analysis of today's media performance. The mainstream media need serious critical feedback, which should be an essential part of the mission of the nation's academic institutions, whose judgments would command respect and receive widespread attention.

Peer pressure among journalists also can have a powerful influence on improving performance. By general consensus, the news media did an abysmal job of covering the 1988 presidential election campaign between Republican George Bush and Democrat Michael Dukakis. Acknowledging that fact, journalists and news executives from newspapers, newsmagazines, television, and radio came together in many forums around the country—most sponsored by the university-based media study centers—to criticize their own work and figure out how to improve the job they would do for the next election.

NBC News Washington Bureau Chief Tim Russert summed up the news industry's attitude about 1988. "No one was satisfied with the 1988 coverage. The public felt cheated by the emphasis on flag-waving and [prisoner] furloughs rather than on deficits and defense, drugs and education. The press felt manipulated by the staging and scripting of 'photo opportunities.' The politicians felt abused by the attention to verbal slips over substance." Months of continuing self-examination produced noticeable improvements in the media's 1992 presidential campaign coverage. Political ads were analyzed and reported on. Issues were addressed more carefully than they had been in previous years. Photo opportunities and slogans supplied by the campaigns themselves were largely ignored or downplayed. Reporters, editors, and producers found ways to free themselves of the campaigns' media managers and spinmeisters and do more of their own independent digging.

The media's performance and the public's interest in the 1992 election also were helped considerably by the lively, unpredictable grassroots candidacy of Texas independent Ross Perot. The fact that the 1992 presidential campaigns spilled outside the traditional boundary lines of the mainstream news media also helped stimulate the public's interest. The mushrooming political use of electronic bulletin boards, on-line services, fax communications, talk radio, and television call-in shows of every description helped to escalate citizen involvement. Voter participation jumped for the first time in decades.

Beyond election coverage, a few television stations, stung by rising criticism of their gratuitous use of violence and sensational and exaggerated reporting, have started to experiment with what they call "family sensitive" news presentations. Their new approach, which seems to be spreading, is not to suppress bad news but to cover the news with greater intelligence and more respect for the audience's sensibilities.

Still, for most of the media, quality journalism remains in short supply, as columnist David Broder and a growing chorus of his colleagues continue to complain. Too many viewers and readers receive an irresponsible, distorted picture of their society and themselves. The most effective way to change that situation is for the audience itself to begin to fight back and demand better.

REFORMING LIBEL LAW

The courts, too, can play at least a limited role in encouraging more responsible performance by the media while not intruding on their editorial freedom. Libel law reform can be a small but significant step in that direction. Currently, the news media enjoy almost total legal immunity from having to pay for the consequences of their own mistakes. Their carelessness and false reporting bear no price. Public figures who have been unfairly or wrongfully attacked on the air or in the press stand almost no chance of gaining redress through libel or defamation suits. One study shows that only 7 percent of those in public life who bring a libel suit against the press succeed. That record is no tribute to the media's editorial accuracy and sense of responsibility. Instead, it merely reflects the fact that the legal standard of "actual malice" has overpowered the rights of libel victims, preventing them from winning cases they deserve to win. The courts have relieved the media from meaningful accountability for their own irresponsible reporting.

Clearly, the nation has a compelling public interest in giving the press the greatest possible freedom to investigate malfeasance, expose wrongdoing, and report on the performance of public officials. The press should have the leeway to make honest mistakes. And it needs full protection from legal harassment that might mute or discourage criticism of government authorities and public figures. But the nation also has a compelling public interest in fostering responsible, accurate, reliable, and trustworthy reporting. To make sound and reasoned judgments, citizens must rely largely on the accuracy of information they receive from the media. A sure way to discourage inaccuracies and encourage more responsible reporting is to make the press pay for its own mistakes that unfairly damage reputations, even if the victims are government officials or public figures.

Victims who have been mugged by an inaccurate story, including those about whom there should be full and vigorous reporting, ought to have recourse to correct the record and seek redress. Public figures are deprived of that right unless they can succeed in proving what is almost impossible to prove—that the press was driven by actual malice, purposeful recklessness in causing their misfortune. The actual malice test for libeling public officials, established by Justice William Brennan in the case of The New York Times v Sullivan, came about for the best of intentions. The Sullivan case occurred in the midst of the civil rights battles of the 1960s and the Brennan decision enabled The New York Times to escape an onerous, undeservedly large libel judgment for an ad that had been placed by the Committee to Defend Martin Luther King. An Alabama jury had awarded L. B. Sullivan, a local police commissioner, five hundred thousand dollars on the grounds that the ad in The Times contained several "minor factual errors" that damaged his reputation. Throwing out the Alabama libel judgment, Brennan said there needs to be "clear and convincing" proof of actual malice, reckless disregard for the truth, in order for a public official to win a libel verdict. The court also required trial judges to review jury findings of fact and make their own independent assessment of the trial evidence on the malice question. Subsequent cases extended the "actual malice" test not only to public officials but also to anyone in the public eye, including singers, actors, and sports figures. To discourage libel suits, an almost unattainable standard of proof must be met before a well-known person can win a libel verdict.

Justice Brennan's *Sullivan* decision served as an important defense of a free press and a needed barrier against lawsuits that threaten to intimidate the media. But it also has had the effect of immunizing the news media from the consequences of their own mistakes and inaccuracies. Some urge the courts to go even further. To encourage the press's watchdog function and promote robust investigative reporting, they want to eliminate *all* libel suits by public officials and public figures, thereby making public figures, in effect, libel-proof: The media would have what Justices Hugo L. Black and Arthur Goldberg called in their concurring *Sullivan* opinions an "absolute and unconditional" privilege. Under such a privilege, public officials would be open targets for media attack and powerless to protect themselves and their reputation, no matter how inaccurate, irresponsible, and scurrilous the story.

Those who would deter all libel suits against the media argue that many are politically motivated and brought simply to intimidate or inhibit coverage of unpleasant matters of importance to the electorate. In my experience, however, the pendulum already has swung too far in creating a free market for journalistic carelessness and irresponsibility. We have seen the rise of an utterly irresponsible, scandal-driven, sensationalist tabloid and television press. And of even greater concern, the requirement to prove "actual malice" paradoxically opens dangerous and potentially highly intimidating questions involving journalistic motivation and personal editorial intentions that the courts have no business pursuing. "Malice," Justice Black pointed out, "is an elusive, abstract concept, hard to prove and hard to disprove." The malice test, he said, offers the press only "an evanescent protection." I agree.

I experienced that problem while defending a libel suit that was brought against NBC in the 1980s by singer Wayne Newton. The Newton libel case was tried in Las Vegas while I was the NBC News president. Newton asked for millions in damages for an *NBC Nightly News* investigative report about his alleged involvement with mob figures, which had aired before I had arrived at NBC News. The report charged that Newton bought an interest in a Las Vegas hotel with the help of mob money. It turned out, according to the singer, that his secret meetings with mob figures had to do with seeking protection against a threat to harm a member of his family, not with financing his hotel interest. To meet the libel case's "actual malice" requirement imposed by *Sullivan*, Newton's lawyers sought to prove not only that the NBC exposé was false but also that it had been motivated by the network's determination "to get Newton" because of some imagined antagonism toward him by NBC entertainment executives. At the trial, testimony was taken about the editorial intent and personal motivations of the NBC News report's correspondent and producer, highly subjective issues and questionable lines of inquiry at best. Probing what is on a journalist's mind and trying to divine his or her motivations in reporting a story is a dangerous game.

In libel trials, motivation should not be the determining factor. In the Newton case, the only test should have been whether Newton could show that NBC's story about him was factually wrong (which it may have been), and if it was, whether he suffered damages as a result. Fortunately, in this country, truth *is* a defense against libel and the plaintiff bears the full burden of proving that the story was false. If the plaintiff meets that burden, the defendant should lose, regardless of motive or malice.

The hometown Las Vegas jury found against NBC and for Newton. It awarded him a verdict that amounted to more than $20 million before the case finally was thrown out on appeal, thanks to *Sullivan*. The amount far exceeded any actual damages Newton might have suffered from NBC's story. Most of the verdict came in the form of punitive damages assessed by the jury against NBC. This was consistent with the pattern of most libel cases against media defendants. The truly dangerous problem that media defendants face is that of excessive punitive damages, which hostile local juries unfortunately tend to impose. Such verdicts can, indeed, be ruinous and even bankrupting to press defendants and can severely intimidate investigative journalism.

To balance the need for a free and unencumbered press, while at the same time encouraging accurate and responsible reporting, the nation's libel laws should be changed to prohibit the award of punitive damages in libel cases. Plaintiffs' recovery, if any, should be limited to requiring a media defendant, who has been found guilty of libel, to publish or air an appropriate correction and pay only the actual damages caused by the offending report. Punitive damages should be eliminated by statute. Perhaps then the courts can be persuaded to abandon the "actual malice" test in libel cases. The media would thus pay for the consequences of their negligence and inaccuracy but not be subject to ruinous penalties. To discourage totally frivolous libel suits, judges should have the power to require that a losing plaintiff pay the defendant's legal costs in such cases.

An evenhanded reform of libel laws that gives litigants a reasonable chance to clear their name against inaccurate or mistaken charges, while freeing the press from the risk of draconian punitive verdicts, may help curb irresponsible reporting and diminish unfounded personal press attacks; at the same time, such a reform might help discourage frivolous and intimidating lawsuits.

In the next few years, libel law issues involving computer bulletin boards and other computer services will be a significant influence in determining the freedom of these services from content domination by their owners. Libel actions have already been filed against operators of on-line services seeking damages for what has been posted by subscribers. The outcome of these cases will help determine the freedom of individuals to exercise free speech on the Internet. Free expression and the multiplicity of voices will be enhanced to the extent that the major corporate operators of on-line services are required (as phone companies, telegraph, and mail services are), to disavow responsibility for the content they transmit.

COUNTERACTING THE
SOUND BITE SYNDROME

Despite their bad reputation, the use of sound bites in news reports is not inherently bad journalism unless they are misused or overused. For people distracted by many different interests and concerns, a news sound bite, like a print headline, can effectively capture the essence of an idea or point of view and help make a news item clear, memorable, and easy to absorb. A sound bite, like a headline, is no substitute for a full story. But merely adding more words does not necessarily add more meaning or depth. A good sound bite stands as a symbol, an icon, often a mnemonic, that gets to the heart of a message. "Read my lips. No new taxes!" was President Bush's campaign sound bite that made his election promise crystal clear. In the end, it helped do him in precisely because voters remembered his vivid, unequivocal pledge and refused to forgive him for backtracking on it after becoming president.

Modern techniques of videotape editing allow for amazingly precise electronic excisions that can be made under extreme deadline pressure without necessarily eliminating or distorting meaning. That was never possible in the days of film and live television, before the invention of videotape and electronic editing. Hence, edited excerpts in the form of sound bites have gotten shorter and crisper, not because news standards have deteriorated but because technology has radically improved. In response to the heavy criticism of television's shallow "sound bite reporting," CBS News pledged at the start of the 1992 presidential campaign to use no short sound bites during its election coverage, a pledge CBS News quickly abandoned. Its use of longer excerpts did little to add substance to or improve the audience's understanding of the CBS News reports. The longer excerpts merely took up more time and made for flabby news scripts.

I spent the first part of my professional life writing advertising copy and producing radio and television spots in an environment in which media advertising space and time were extraordinarily expensive. In advertising, brevity is an essential ingredient. To this day, I marvel at just how much useful information (and misinformation as well) can effectively be contained in a single succinct line of advertising copy or a pithy ten-second sound bite, and how much of it the brain can absorb and retain. True, sound bites usually eliminate nuance and detail. But at times that can serve as an advantage. Their sheer simplicity and unequivocal clarity tend to sharpen focus, add intensity, and contrast differences.

Nothing but sound bites, however, is a poor substitute for a coherent story and full explanation. A total diet of sound bite and headline news leaves a disproportionate sense of exaggerated and disconnected crises and jumbled emergencies. It also leaves out the perspective and background that are needed to gain in-depth understanding and comprehension. To overcome the sound bite syndrome, a sustained effort should be made to increase the opportunities for extended live appearances by major candidates and public officials on prime time news, public affairs, and talk shows. This is not as naive or unrealistic a proposal as it may appear in the current "dumbing down" competitive climate.

Nobody can force the networks' flagship news shows—which still attract the biggest nationwide news audiences—to open up larger chunks of their time for important public discussions and substantive newsmaker appearances. But the network nightly news broadcasts are an endangered species. They are losing audience share, and network affiliates have steadily been pushing them back into less desirable time periods. In the interest of their own competitive survival, the network news half hours should modify their age-old formats to spend more time reporting, examining, and analyzing the day's most important news stories. Fewer items, more depth of coverage, and more seriousness of purpose would give those daily network reports a new life and distinctive character in an increasingly duplicative and superficial headline news environment. I am convinced it would be in their own self-interest to go against the grain and upgrade the quality of their news presentations, rather than continue to head down-market and sink into the muck of local and syndicated pseudonews and "reality TV."

Today, CNN, C-SPAN, and public television provide much longer news coverage than network or local television traditionally have offered, including the televising of entire campaign "stump speeches" and full debates. Obviously, news and public affairs coverage that is not confined exclusively to news headlines and staccato bits and pieces and that has room for extended discussion can be enormously helpful in improving understanding for the motivated viewer or reader.

By contrast with a sound bite, whose pithiness can disguise the absence of substance, one benefit of extended coverage is its capacity to reveal ignorance, evasion, and superficiality. Given more air and press time, such characteristics become more difficult to hide. Ross Perot sounded much more interesting in sound bites, spouting aphorisms and colorful phrases during the 1992 campaign, than he did when put in the position where he had to answer reporters' questions and explain his policies and programs at length. Perot's obvious limitations in his knowledge of the specifics of his tax and other proposals during an early *Meet the Press* interview caused him to go into seclusion for a period during the campaign to do more homework. Despite Perot's frequent complaints that the media never gave him enough time to discuss his ideas in full, whenever he was given the time he seemed to come up wanting.

During presidential election campaigns the network news broadcasts should televise extended and frequent interviews with the major presidential and vice presidential candidates. During the climactic final month of the campaign, such live candidate interviews should appear weekly on all the networks. And between elections, the network nightly news half hours should pursue a similar strategy, dealing in depth and at length with the major issues as they arise. The more free time candidates and issues receive on television news and public affairs programs, the less influential paid political advertising will be in the battle for public support.

One of the virtues of the 1992 presidential campaign coverage was the frequent, uninterrupted direct exposure of the major candidates on television and radio just about everywhere *except* on the network evening news shows. Contrary to conventional wisdom, I am convinced that the more substantive the nightly network news shows become, the better they will fare in the ratings wars.

IMPROVING THE QUALITY
AND DISSEMINATION
OF CIVIC INFORMATION

Shortly before his death, James Madison wrote to a friend, "A people who mean to be their own governors must arm themselves with the power which knowledge gives." In the electronic republic, the American people increasingly will "be their own governors." Yet, under the constraints of the First Amendment, government itself is limited in what it can do to dispense useful political and civic information to them. Essentially, that job is left to the media and to a wide and disparate assortment of political advocates, special interests, public service groups, candidates, political parties, corporations, labor unions, and others. But even within the limitations imposed by the First Amendment, the public sector should have a vital role to play in providing useful and essential political and civic information to all citizens. After all, government still has the primary responsibility for educating the people, both inside and outside the classroom. And what could be more important than educating the people for responsible citizenship?

To promote democracy throughout the world, the United States Information Agency, the Agency for International Development, and a host of other federal agencies spend almost a billion dollars a year. To promote their own activities, thousands of government departments spend additional hundreds of millions of dollars on publicity, public information, and press relations in this country. Yet federal, state, and local governments do almost nothing to create and distribute critically needed civic information to their own constituents and to promote active, informed democratic participation among all of the citizens of the United States.

The time has come to change that. The Library of Congress in Washington, D.C., for example, provides a wealth of solid and useful information, data, and research services about public issues and legislative matters to members of Congress and their staffs. The library's legislative reference service is entirely nonpartisan, universally respected on both sides of the aisle, and essential to the work of the Congress and the executive branch. In the information age, in which knowledge is power, what better resources could be made available to the public at large than the exceptional legislative information services of the Library of Congress?

The nation's taxpayers, after all, support the library's existence. Who could do a better job of carrying out the essential task of educating and informing the nation's citizens with useful, responsible facts and information about the tough, complex issues that confront the nation? In the electronic republic, in which public opinion often will be "the final determinant of economic and political action," the public needs to be informed every bit as much as its elected officials.

The Library of Congress should be given the budget to expand its mission to educate not only the Congress and the executive branch but also the nation's electorate. In the summer of 1994, the library announced its intention to transfer its vast collection of published works into digital electronic data streams capable of being delivered at the speed of light anywhere in the nation. That same capacity should be applied to the library's legislative reference services.

The civic information the Library of Congress produces for the public at large needs to be clearly and interestingly written, attractively designed, and widely disseminated in a full range of print and electronic formats—including fax, audio- and videocassettes, CD-ROM, modem, the mails, and on-line services. The library should seek out the widest possible audience of citizens for the information it distributes about important issues and public policy matters. Every citizen and every public library, school, and educational institution that requests the help of the Library of Congress in disseminating information about the public sphere should receive whatever it needs on reasonable terms.

The political parties are another public institution that should be instrumental in providing useful political information to the nation's citizens. The parties need to return to their roots to help educate and inform their constituents about political matters. Once the parties were the chief catalysts for grassroots political action and, even more important, the major source of political information to citizens in every community. In the parties' understandable zeal to raise money for their increasingly expensive, high-tech election campaigns, they have all but abandoned their traditional role as the primary political and civic educators of the electorate.

It seems only logical to urge the parties to revitalize their efforts to educate and inform their constituents about the key issues of the day. Of course, what they provide will be partisan information. But the parties' efforts would add importantly to the public dialogue not only during election campaigns but also during the periods between elections, when the major public issues actually get debated and decided both locally and nationally.

DEALING WITH NEGATIVE ADS AND PERSONAL ATTACK CAMPAIGNS

During elections, negative advertising and personal attack campaigns have come to dominate and, many feel, blight the entire election process. Misrepresentation, oversimplification, exaggeration, and just plain nastiness abound, creating an ugly political environment that triggers massive public hostility, distrust, and frustration. As election day approaches, the frequency and intensity of these mean spirited ads increase; they become ubiquitous and inescapable, polluting the atmosphere across the nation. What steps should be taken to diminish their presence and counteract their corrosive and destructive force?

The reason attack commercials and negative campaigns have become ubiquitous is that they consistently register more impact on the electorate than affirmative and constructive campaign messages that focus on substantive issues. Research shows that the nastier the campaign commercial, the more memorable and therefore the more effective it is. In addition, since serious issues often are avoided in campaigns because candidates do not like to say what many voters may not want to hear, most of what a candidate has to sell is his or her personal character, and that is what the negative ads try to undercut. Negative attacks produce results. So, despite escalating criticism of the genre and admirable intentions initially to refrain from trashing one's opponent, attack commercials invariably surface when the battle grows intense, the race gets too close for comfort, and time starts to run out on a candidate determined to do whatever it takes to win.

The United States is one of the very few countries in the world that permit paid political advertising. In recent years, numerous efforts have been made to ban or limit negative attack commercials and even to outlaw political advertising altogether. Nobody that I am aware of has proposed to so limit or censor negative print ads, direct mail campaigns, audio- or videocassettes, or on-line computer bulletin boards or dialogues. But federal legislation has been proposed that would require candidates whose television commercials attack their opponents to appear personally in those commercials. Another proposal would require television stations that accept such commercials to offer free airtime for the attacked candidate to respond.

In a society that cherishes free speech, none of these prohibitions would appear to be either feasible or especially in the public interest. Tempting though it may be to put an end to the offensive commercials that pollute elections, the proposed cures set dangerous precedents. Their effect would be to censor the conduct of political campaigns by restricting what can be said about opponents and issues. If negative paid political ads are banned, should other attacks—in letters, direct mail, computer bulletin boards, or posters—also be banned?

Such proposed restrictions open a Pandora's box of dangerous government regulations that would require questionable definitions about what constitutes a "personal attack" or a "negative statement." They would sanction intrusions by public administrators into the free flow of ideas at the worst of all times, in the midst of an election campaign. Imagine a government official trying to decide in a heated campaign whether a commercial is negative and should be banned, or whether it makes a legitimate point that is appropriate and should be allowed to air. Nothing would be more chilling or inappropriate than to have a government agency cast as a national watchdog over election campaign messages. As distasteful and even destructive as political campaigns get, the integrity of the election process requires that their content be free from government meddling. The electorate in the voting booth, not a government official, should decide whether a campaign or a candidate has gone too far and behaved irresponsibly.

That is not to say, however, that absolutely nothing can be done about nasty attack ads. Candidates who accept public election financing should be required to vouch personally in specific public statements for each of their own campaign commercials and messages. Taxpayers help pay for the production of these materials. They are entitled to know that the candidates who receive taxpayer funds take personal responsibility for what their campaigns are saying. Public disclosure, which imposes no censorship and entails no government-enforced content limitations, is one step that can be taken. Public disclosure would promote a measure of personal responsibility for each candidate's political advertising.

But what about campaign materials produced independently by those who may be supporters of a candidate but are not part of his or her election team? The notorious Willie Horton commercial involving a black prisoner who committed rape while on parole, which was used in support of Bush during the 1988 presidential campaign, was produced and paid for by a group that claimed to be totally independent of the Bush campaign. Purchasers of paid political advertising already are required by law to identify themselves in the ads and commercials they buy. It would be a logical extension of that disclosure principle to require candidates who accept public campaign funds to personally endorse or repudiate all advertising bought on their behalf, even ads produced by so-called independent groups. At the very least, such a requirement might deter candidates from relying on surrogates to do their dirty work for them.

Putting candidates on the record endorsing or repudiating negative attack ads will give the electorate useful information in deciding who will get their votes. Taking that step will not satisfy those who want to suppress either negative advertising or all political advertising. But as columnist David Broder wrote, "Such accountability requirements might make the candidates think twice about what they're putting on the air—and maybe clear the air of the worst of the pollution."

IMPROVING PRESIDENTIAL DEBATES

For almost two decades now, televised debates have been the centerpiece of every presidential election campaign. High drama happens when the major presidential candidates are pitted against each other on a single platform before a nationwide audience. The appeal of presidential debates is irresistible, as has been demonstrated by their record-breaking audience ratings.

Debates also have their limits, of course. It takes more than superior debating skills and a pleasing television manner to make a presidential nominee into a good president. Still, as the ratings and surveys suggest, voters view the presidential debates as critically important events. But probably because the stakes are so great, most presidential debates, like most Super Bowls, have not lived up to the public's high expectations for them. Many have lacked both substance and spontaneity. Part of the reason for this may be that presidential debates invariably have been difficult to arrange, their formats hard to obtain agreement on. Front-running candidates, especially incumbents, consistently exercise more control over the debate formats and arrangements than is either fair or appropriate. For the candidate who is ahead in the polls, a poor debate performance can reverse his or her campaign momentum and wreck the chances for election. For the candidate who is behind, a winning performance or an opponent's blunder in a televised debate offers an instant opportunity to overtake the front-runner.

With the prospect that the entire election might well ride on the outcome, most presidential debates have been carefully choreographed pseudo-events, staged appearances rather than spontaneous and revealing discussions. They reveal too little and disguise too much to be of substantive use to voters in the audience. Presidents Johnson and Nixon saw to it that no debates took place with their presidential rivals during their election campaigns. That almost happened again in 1988 when the polls indicated that George Bush's narrow lead over Michael Dukakis had suddenly increased dramatically. James Baker, Bush's campaign manager and debate negotiator, threatened to have his candidate boycott the debates unless his one-sided conditions for their format and scheduling were met. In local political races, that happens all the time. Candidates find it expedient to avoid debates altogether if the polls show they are ahead, confining themselves instead to controlled campaign events, paid commercials, and scripted appearances. (Sometimes that tactic can backfire, as happened in the 1994 New York gubernatorial campaign. Incumbent Mario Cuomo refused to debate his relatively unknown Republican rival George Pataki, who ended up winning the election by a narrow margin.)

In 1992, the presidential campaign's second Bush-Clinton debate broke new ground when, under the direction of the Presidential Debate Commission, the usual political journalist questioners were replaced with ordinary citizens in the studio audience. By the time the second debate had ended, it was clear that the citizens at large had done a better job of raising issues that were of concern to the electorate than the professional journalists had ever accomplished. The quality of the presidential debates improved substantially.

Notwithstanding the emphasis that debates place on who's ahead, who's behind, who's gaining ground, and who's losing it, televised political debates remain an essential tool for informed public decision making. They should be required for every presidential campaign, as well as for other important campaigns for public office—local, state, and national—that receive matching taxpayer campaign funds. Candidates who receive such funds should be required to commit to participating in televised election campaign debates. At the very least, that up-front commitment should help reduce the one-sided control that incumbents and front-running candidates usually exert in the debate process.

In recent sessions of Congress, legislation has been introduced in efforts to make the presidential debates more substantial and less subject to political manipulation. One proposal would require that all presidential debates be produced not by the political parties but by an independent, nonpartisan organization such as the League of Women Voters. (However, dissatisfaction with the League's record of organizing earlier presidential debates led to the formation in 1988 of the Presidential Debate Commission, headed by former political party chairmen, to oversee the job.) Another bill proposed to improve the format of the debates by specifying that a minimum of thirty minutes be reserved each time for a direct confrontation between the presidential rivals.

Such congressional micromanagement of debate formats is both poor public policy and makes little practical sense. But the threshold requirement, that candidates who accept matching public campaign funds must commit to debate their opponents, should, in my opinion, be adopted. That proposal has vigorous dissenters, among them, for example, ABC political analyst Jeff Greenfield. The dissenters argue that forcing candidates who accept public matching funds to debate would infringe on their First Amendment right to speak whenever and wherever they please, or keep silent if they prefer. The goal, however, is not to limit, censor, or regulate campaign debates, but to encourage them to happen. The proposed requirement to debate is consistent with the essential purpose of public campaign financing—to promote substantive political dialogue among citizens, not to inhibit or reduce such dialogue.

THE EXCESSIVE INFLUENCE AND UNBALANCED USE OF MONEY IN POLITICS

The money issue becomes especially acute in an electronic republic, where the public plays a central role in the day-to-day decisions of government. After all, money is the key ingredient in being able to reach and influence public opinion. How should the sharply escalating costs not only of election campaigning but also of all political advocacy be dealt with in the telecommunications age?

As we have seen, politics has become a prohibitively expensive battleground in which access to voters is available disproportionately to those with the most money to spend. For example, corporations with major interests at stake are those that can most readily afford to buy access to the media and to conduct the most expensive publicity campaigns to get their political point of view across.

A study of fund-raising patterns by the California Commission on Campaign Financing found that in the eighteen highest spending initiative campaigns, 69 percent of all contributions came in amounts exceeding one hundred thousand dollars. How can the increasingly distorting effects of money on political decision making be dealt with in the electronic republic? Is it possible to overcome the growing disparity in resources between moneyed interests that can dominate the debate and public interest advocates who can barely be heard over the political din?

Urgent steps must be taken to eliminate the undue and unfair influence of large contributions, enhance participation by small donors and nondonors, limit campaign spending, put caps on funds available for lobbying, and in general place political deliberation under some measure of fair and appropriate fiscal restraint. Efforts to fashion effective and constitutional restraints on political spending may be difficult, but they are not impossible. To expect to achieve a totally level playing field for all sides in all elections and for all political advocacy would be unrealistic. One side will always have more resources at its disposal than another, or will enjoy more public support, or have more energetic and committed backers, or richer ones, or better fund-raisers. But the goal must be to ensure that the public has at least enough access to information about all sides of any important public issue to make reasoned decisions. And the costs to achieve a reasonably fair hearing should not be so onerous as to preclude any responsible point of view from getting appropriate exposure.

In the interest of political integrity, PAC donations should be severely limited if not eliminated altogether, and personal political donations should be capped. Contributions of so-called soft political money should be strictly controlled and well publicized. (Soft money is the term used to describe payments to political parties to promote public participation and turn out the vote. Such payments have not been subject to the same federal accountability requirements as outright campaign donations and are frequently used to circumvent such requirements.) There should be matching public campaign financing for all federal elections and, wherever possible, for elections at the state level as well. As a condition of accepting such financing, limits should be placed on campaign expenditures, including closing the soft money loophole. Extensive free time should be made available for political advocacy on all public telecommunications facilities. And, as already discussed, commercial media, both print and electronic, should be encouraged (but not required) to open up free space and extensive time for advocacy during election campaigns as well.

Tax deductions for corporate lobbying and corporate political advocacy, even if directly business-related, should be eliminated. Citizens get no tax deductions for their own political efforts. The public should not be put in the position of actually subsidizing corporate political spending. What the Supreme Court, in another context, called "the corrosive and distorting effects of immense aggregations of wealth that are accumulated with the help of the corporate form" must be minimized. Corporate wealth should no longer be permitted to "unfairly influence elections." Overall, the objective should not be to achieve spending parity by all sides in political campaigns; it should be to make sure that all sides have a reasonable opportunity to reach the public with their positions and see to it that the onerous costs of political action are reduced.

11

"TO INFORM
THEIR DISCRETION"

In 1994, the year British poet Stephen Spender turned eighty-five, he wrote a new introduction to his autobiography, *World within World,* first published in 1951 when Spender was only forty. The Spender autobiography, which was reissued after being out of print for a dozen years, vividly chronicled the period between the two world wars. Spender called his introduction to the new edition "Looking Back from 1994," and in it he described "the great difference between Then and Now":

"[T]he street, the city, is not the reality as it was in the Thirties with Hunger Marches and great demonstrations. Today we see the unemployed, for the most part, on television screens. They are mirror images moving across glass, collected for us from all over the world by camera crews and some reporter who interprets them for us. These are three-dimensional figures transformed almost to abstractions— while we sit in our rooms perhaps eating our dinners or high teas, and watch them. The viewer may be shaken or depressed or distressed by these images but they do not step out of the screen into his room, or shout from the street up to his windows. Today we have become spectators of reality, which has become a photograph."

Spender was right, of course, that television screens, showing "mirror images moving across glass, collected for us from all over the world," have turned us into mere "spectators of reality." But he did not go far enough. Had he described "the great difference between Then and Now" merely a year later, Spender might also have observed how remarkable new interactive telecommunications technologies now enable television viewers to move from "spectators of reality" to active participants. The viewers of television's images could, in effect, now "step out of the screen" and, acting as more than spectators, respond to them. The television screen, when connected to a personal computer and a modem, can empower Spender's "distressed" viewer to act. Direct-dial, Touch-Tone, and cellular telephones; audio- and videocassette recorders; Minicams and personal computers; desktop printers and photocopiers; fax machines and modems; personal computers and computer networks; electronic bulletin boards and microprocessors; telecomputers and satellite dishes; broadband networks and fiberglass wiring; talk radio, cable television, and express mail . . . : All these and more have created direct electronic links between individual citizens and city hall, the county seat, the state house, and Washington, D.C.—especially between individual citizens and Washington, D.C.

As one *New York Times* Washington reporter observed, "The cumulative effect has been to turn a somewhat slow and contemplative [political] system into something more like a 500-channel democracy, with the clicker grasped tightly in the hands of the electorate." While only a small fraction of the public is yet wired into the new interactive electronic media, the amplified messages of those who are connected pack a considerable wallop. And polls, focus group interviews, telephone samplings, and escalating use of ballot initiatives and referenda give voice to the rest. The people at large either have taken into their own hands the power to decide laws and public policies, or they powerfully influence those in the executive, legislative, and even judicial branches whose job it has been to make such decisions.

Ironically, at the same time we see the political process open up to the citizenry and change the balance of power in our increasingly interconnected and wired society, we also see an epidemic of public disaffection and frustration with politics. Except for the 1992 presidential election, citizens' political participation has declined, not expanded. People perceive that government has failed to represent their interests; that citizens have little meaningful say; that, notwithstanding sunshine laws, what politicians do is largely hidden from public view. Back in 1964, University of Michigan pollsters asked people, "How much of the time do you trust the Government in Washington to do what's right?" Seventy-six percent said most or all of the time. Twenty-eight years later, when asked the same question, only 29 percent said they trust the federal government most or all of the time.

On the one hand we see widespread distress and despair over political gridlock, hidden political agendas, and unresponsive politicians who live "in splendid isolation" and are dominated by special interests. On the other hand we see equally widespread concern that ordinary citizens are uninterested and not sufficiently prepared to carry out their responsibilities of citizenship.

The fact is, active and responsible citizens are made not born. Good citizenship must be cultivated and cannot be left entirely to chance, as is currently the practice in American society. The ancient Greeks invested major resources and much time and effort to encourage what they called the public spirit in their republics. We should follow their example. As the American political system transforms itself into the electronic republic, we need to:

1. Restore civics education for all grades in the nation's classrooms. Americans should be taught from the earliest age the fundamental requirements and responsibilities of citizenship and the importance of fulfilling their civic responsibilities.
2. Dramatically improve the quality, appeal, and dissemination of solid, responsible information about news and public affairs, both in print and electronics.
3. Reshape the nation's civic and political institutions to take full advantage of the opportunities afforded by new communications technologies.
4. Reshape current political processes that function poorly, are outmoded, and contribute to the public's growing frustration with politics.

THE NATURE OF RESPONSIBLE CITIZENSHIP

The basic principles of what must be done are hardly new or original. Almost twenty-five hundred years ago, Aristotle made the point that "democracy . . . will be best attained when all persons alike share in the government to the utmost." More than a century ago, Tocqueville echoed that insight: "perhaps the only means that we still possess of interesting men in the welfare of their country," he said, "is to make them partakers in the government."

Tocqueville was impressed in the early nineteenth century by the prominent role that politics played in the personal lives of so many of the Americans he observed. He viewed the American people's dedication to public affairs as the main reason for the remarkable vitality of American democracy and its strong responsiveness to public concerns. "To take a hand in the regulation of society and to discuss it is [the average American's] biggest concern and, so to speak, the only pleasure an American knows," Tocqueville wrote. "This feeling pervades the most trifling habits of life; even the women frequently attend public meetings and listen to political harangues as a recreation from their household labors. Debating clubs are, to a certain extent, a substitute for theatrical entertainments. . . . [I]f an American were condemned to confine his activity to his own affairs, he would be robbed of one half of his existence; he would feel an immense void in his life which he is accustomed to lead, and his wretchedness would be unbearable. . . . The humblest individual who cooperates in the government of society acquires a certain degree of self-respect."

Many democratic political theorists have argued that active and widespread public participation in civic affairs enriches the lives of those who participate and that civic involvement affords people a lofty and satisfying sense of purpose and self-fulfillment. Today, the idea that engagement in public life offers a kind of nobility and higher calling—first articulated in modern times by John Stuart Mill—seems to have gone out of fashion in many quarters. The ideal of dedication to public service has been displaced to a great extent by the pursuit of private interest.

American liberal philosopher John Dewey provided the contemporary rationale for American society to view widespread civic participation and collective deliberation as high priorities. As in jury trials, truth has a better chance of emerging from group deliberation than from a single authority, Dewey said, no matter how expert, experienced, and sophisticated that authority may be. Public decision making, he argued, achieves more than the sum of its parts; nothing is more important than finding new ways to create an "organized, articulate Public." He insisted that "inquiry and communication" are the keys to a functioning democracy.

Building on that foundation, political theorist Hannah Arendt proposed a new organizing structure for modern democratic government, which she called "the council state." Like ancient Greek democracy, the council state would be rooted in active and widespread citizen participation, which, according to Arendt, is the essence of political life. "The councils say: 'We want to participate, we want to debate, we want to make our voices heard in public, and we want to have a possibility to determine the political course of our country.' " For Arendt, the essence of democratic politics is the process of interacting, persuading, and negotiating in a common cause—not merely voting on election day in a booth, which "has room for only one." In the days before the development of interactive electronic communications, Arendt proposed that the United States be organized into concentric circles of citizens' councils, whose deliberations and participatory decision making would radiate out in waves to encompass the entire electorate.

In recent years, University of Texas Professor James S. Fishkin has developed a specific structure, a practical application of Arendt's citizen council idea, which he calls "the deliberative poll." Fishkin argues that typical random polling of public opinion, which is so prevalent in today's democratic governance, is deeply flawed. People tend to give ill-considered, snap answers to public opinion polls, according to Fishkin, simply to satisfy the pollster's need for a response, even though the respondents may be uninformed or poorly informed about the specific issue being addressed. Thus, poll results reflect nothing but a quick, undeliberated consensus at a particular moment in time. By contrast, for Fishkin's "deliberative poll," random groups of people would be carefully chosen to reflect the characteristics of the entire population. A chosen group would be invited to spend two or three days at a model citizens' convention to thoroughly consider and debate a particular issue. Experts would put the case for and against. The survey group, representing all citizens, would raise questions and discuss the issue intensively among themselves. At the end of their deliberations, the convention would be polled and the citizens' responses then presumably would accurately reflect how the nation as a whole would vote if all citizens had been so thoroughly informed and given an equal opportunity to deliberate with each other about the issue in question.

In another variation, Professor Robert A. Dahl suggested the creation of a "minipopulus," perhaps 1,000 citizens randomly selected to deliberate for a period of time on an issue. The minipopulus could hold its meetings on-line through telecommunications and then announce its recommendations, which would be distributed electronically to all citizens. Different minipopulus groups might discuss different major issues, according to an agenda laid out by the Congress together with the president. Each minipopulus would be served by experts, specialists, and scholars, as well as a core administrative staff.

Hearings might be held for the minipopulus. The judgment of the minipopulus groups would tend to represent the judgment of all citizens, but would be advisory only.

Thus new mechanisms would be tried to adapt to the new conditions of the democratic process, transforming the system to suit the new era of the electronic republic. The minipopulus would not replace legislative bodies, but supplement them. As Professor Dahl pointed out, "the democracy of the future will be different from the democracy of the past." It will never be perfect, ideal, or complete. Democracy's exacting requirements will never be fully met. But the vision of people governing themselves as political equals, and possessing the resources and institutions necessary to do so will always be with us.

We should not, however, make too much out of the findings of deliberative polls or the conclusions of samplings of citizens chosen by social scientists to reflect the population as a whole. Council-states and deliberative polls make for interesting political and social experiments and perhaps even informative television shows. But they are limited by their strong sense of artificiality and lack of real-life application. They seem hardly suitable mechanisms to be influential in governing a large and sophisticated modern nation-state in which no artificially selected sample group can effectively stand for the whole, especially when deciding major matters of public policy. These efforts, however, do reflect the yearning to bring participatory democracy down to manageable human dimension, in which the public at large can participate directly in deciding the policies that affect their lives.

The question is whether, in this age of interactive telecommunications, we can take advantage of the new technologies to help make modern deliberative democracy work better than it does today. Nobody yet knows for sure. Too few people are yet interconnected by means of computers; interactive technologies and skills are still only in their infancy. But much can be done even with today's relatively primitive interactive technology (primitive compared only to what we can expect in the decade ahead). Interactive broadband transmission capacity is expanding on many levels, adding video to the plain old telephone and adding interactive telephony to plain, not so old cable television. Yet the United States is doing precious little in a serious and systematic way to explore how these new interactive communications systems can be designed to enhance the local and national democratic process. The information superhighway is being shaped by random forces for other purposes in a fluid marketplace. While we lament the fact that the United States lacks effective political parties or other mechanisms to mobilize the participation of the average person in politics, unlike the ancient Greeks, we have no continuing, nationally organized effort to promote widespread public participation in the affairs of government.

Television and radio have made it possible for citizens to attend political events and see and hear their political leaders more often and more directly than ever before, even if they are only in the form of what Spender described as "three-dimensional figures transformed almost to abstractions." But the audience at home who watches the glass screen is, by definition, isolated, autonomous, and private—the antithesis of the nation's traditional robust politics, which in the past served as a catalyst to bring people together to listen, learn, talk, share opinions, and be entertained. The 1992 election began to restore that old-time feeling as presidential politics at last broke out of the confining ghetto of the news media and moved into a broader realm of public life. We need to do more of that in the years ahead.

THE FEDERAL COMMISSION ON CITIZENSHIP

In the electronic republic, the cultivation of effective citizenship must be made a clear national priority and a vital and continuing element of public policy. In view of the low state of the public dialogue and the declining conditions of effective citizenship, the time has come to inaugurate a broad-scale national effort to improve the quality of civic engagement. We could do worse than emulate the ancient Athenians.

The president should appoint, and the Congress confirm, a highly visible Federal Commission on Citizenship to prepare the people of the United States to exercise their civic responsibilities for the twenty-first century. The commission, with members from every state, should be charged with a broad mandate to recommend basic reforms in our political processes and institutions and to stimulate widespread public participation in the nation's civic affairs. Its focus should be local as well as national. Funds for the commission's work should come from both public and private sources.

Several precedents for such an initiative come to mind. The earliest, in many ways the most useful, and certainly the most intriguing is the Athenian Council of 500 that operated during the period of direct democracy in ancient Greece. The council, as we have seen, had the chief responsibility to prepare the citizens of Athens to deal with the issues that came before them. The Council of 500 circulated information, organized and conducted preliminary citizen discussions, prepared the agendas, and framed proposals for the public assemblies on the hill of Pnyx to consider. The council had no decision-making power of its own. Its members were chosen by lot. They were paid for their efforts and were limited to two yearlong terms. Over the years, therefore, the council's responsibilities were spread among a large number of the citizens of Athens. The work of the Council of 500 was considered so important to the governance of the city-state that a tenth of its members, the so-called *prytaneis,* were required to be in constant attendance in the marketplace to consult with Athenian citizens who sought them out there. Some members of the council even were assigned to sleep in the marketplace.

As we've discussed in chapter 2, the ancient Greeks considered it essential that all citizens become deeply involved in civic affairs. The community worked hard to make that happen. In Plato's *Republic,* political knowledge was considered the supreme art. Anyone who did not take part in the affairs of state was regarded, as Pericles put it, "not as someone who minds his own business but as useless." Family members and fellow Athenians considered it their duty to educate each other about citizenship and promote widespread public engagement in civic activities. Looking toward the electronic republic in twenty-first century, the United States would do well to model its Federal Commission on Citizenship on the ancient Athenian Council of 500.

A more contemporary model for the Federal Citizenship Commission was President Reagan's Grace Commission, established by executive order in 1982 to examine ways to eliminate waste in government. Headed by businessman J. Peter Grace, the commission was funded and staffed by the private sector. Its work cost seventy-five million dollars and took eighteen months to complete. Thirty-six task force teams chaired by people from corporations, the academic world, and labor, and staffed by volunteers, looked into every department of government for sources of waste and opportunities to make economies.

For the Federal Citizenship Commission, task force teams should be organized to examine every current political institution and process, including the parties, election financing, initiatives and referenda, civic education, information dissemination, and voting. Its recommendations for improvement and reform should be designed specifically to be suitable for the telecommunications age.

A third model to examine would be the Commission on Presidential Debates, organized in 1987 to improve and, if possible, guarantee the continuation of presidential campaign debates. As president of PBS and NBC News, I participated in two studies of the presidential election process in the 1980s that led to the establishment of that commission. The first, chaired by former FCC Chairman Minow under the auspices of the Kennedy School of Government at Harvard and sponsored by the Twentieth Century Fund, focused specifically on presidential debates. It sought ways to improve the debate formats and, more important, to figure out how to ensure that presidential debates, absent from the election campaigns of 1964, 1968, and 1972, would take place every four years.

The second study, led by two wily and influential political veterans, Democrat Robert Strauss and Republican Melvin Laird, had broader, more ambitious goals. It sought to reform a number of deficiencies in the presidential election process that had become evident in the television era. Prominent among these was the presidential primary system, whose importance in picking the nominees had mushroomed in a single decade but whose haphazard and wholly uncoordinated structure raised numerous questions about its viability. The Laird-Strauss task force also studied the presidential debates; the increasingly corrosive phenomenon of negative attack advertising; and the question of how to elevate the public's interest in key political issues, in contrast to the growing emphasis on the private lives of candidates.

The National Commission on Presidential Debates, chaired by former leaders of both political parties, was the single most significant practical result of both of these studies. In 1988 and 1992, the commission produced the presidential and vice presidential debates, the most important and most watched events of both campaigns. The debate commission's apparent success in assuring that presidential debates would be part of future election campaigns offers a useful prototype for the Federal Citizenship Commission's far more complex task of reawakening civic consciousness, stimulating citizen involvement in public affairs, and reforming political processes.

Still another interesting, if rather unorthodox source that should be looked to as a model for at least some of the citizenship commission's work was one established in 1993 by the Russian parliament as a kind of public ombudsman group to monitor the fairness of the parliamentary elections after the downfall of Communism. Originally bearing the unlikely name of the "Tertiary Information Court," it worked sufficiently well for the Russians to decide to continue the court on a permanent basis under a new (and equally unpromising) name, the "Russian Federation Presidential Judicial Chamber for the Adjudication of Information Disputes." The court, now chamber, was put in place to have moral and ethical but not legal authority over the conduct of the Russian parliamentary election campaigns as part of the Russians' new experiment with democracy. The chamber's members, appointed by parliament, are charged with the obligation to be independent, nonpartisan, and fair to all factions. Its role in the 1993 election, as the chief judge described it at a conference I attended in Moscow in June 1994, was to make findings and recommendations regarding the fairness of the parliamentary election process, but not to impose punishment or order changes in what the candidates and political parties could do in their campaigns. The Tertiary Court could only pursue inquiries and publicize its findings in the hope that those whose practices it criticized would feel the pressure to reform themselves. During the 1993 Russian parliamentary election, for example, the court urged candidates who worked as television commentators either to take a leave of absence from their television jobs during the campaign or end their candidacy. The court found that the commentator-candidates had taken unfair advantage of their access to television while running for election. In another set of cases, the court criticized incumbents and government officials who, it found, used their appearances on government-controlled television channels to engage in "covert political canvassing." The Tertiary Court's findings and recommendations received widespread media coverage and, according to the chief judge, "the result was that public opinion made the parties and some candidates change their behavior." Other Russians I talked to in Moscow were not quite so enthusiastic, complaining that in the 1993 election the court took too long to issue its findings and ignored too many problems.

In the United States, the Federal Election Commission, whose role is limited to monitoring election spending and public financing, is generally viewed as a relatively timid agency that is susceptible to political pressure, although it did impose heavy fines after the 1992 and 1994 elections for several major campaign financing violations.

Finally, still another prototype for the proposed citizenship commission should be the influential work of the Carnegie Commission on Educational Television, the distinguished fifteen-member panel whose two-year study in the mid-1960s led to the Public Broadcasting Act of 1967 that created public television in the United States. The Carnegie Commission, chaired by Dr. James R. Killian, Jr., of MIT, was privately funded and had the support of President Lyndon B.

Johnson, who viewed public television as one of his most important legislative priorities.

While some of the precedents suggested here have produced more successful results than others, all played significant roles in bringing improvements to the public sector. A comparable Federal Commission on Citizenship, with attributes from both ancient and modern prototypes, would take on the job of making recommendations to modernize the American democratic process and improve the quality of citizenship for the twenty-first century. The commission should take on the task of addressing four major areas that need the most attention: promoting education and training for responsible citizenship; marshaling the new technologies to inform public deliberation; reforming the nation's political processes and institutions for the telecommunications age; and restoring a sense of personal citizen responsibility for the well-being of the general community.

RESTORING CIVIC EDUCATION

Training for responsible citizenship must start at an early age for all the nation's children both at home and in the classroom. A strong commitment to public education has been central to the development of American democracy. "The strength and spring of every free government is the virtue of the people; virtue grows on knowledge, and knowledge on education," wrote Moses Mather in 1775. To the heavy burdens already imposed on the nation's underperforming educational systems must be added the obligation to provide effective lessons in civic responsibility and democratic citizenship. If the commission were to do nothing else but push for ways to enrich and improve education for citizenship, it would make a considerable contribution.

DEVELOPING TECHNIQUES FOR STIMULATING THE PUBLIC DIALOGUE

The people of the United States have a long tradition of direct citizen action and grassroots civic initiative. Today, concerned about the declining public involvement in such affairs, a group of communities as widespread as Salem, Oregon, Wichita, Kansas, and Charleston, South Carolina, have launched impressive initiatives designed to get more citizens engaged in town meetings, civic forums, media exchanges, issues workshops, school programs, and general community activities.

On the federal level, few such public initiatives have been tried, especially when no election looms. It certainly has not been customary for agencies of the federal government to engage citizens directly in developing public policies or to involve them from the beginning in drafting laws and regulations. Typically there is considerable bureaucratic resistance to a bottom up approach to policy making. In the electronic republic such resistance will have to be overcome and new procedures developed to accommodate increased public involvement.

Through new interactive technologies, new opportunities are becoming available to get members of the public engaged in resolving the issues that directly affect them. As we have seen, cable and on-line networks in Santa Monica, California, Reading, Pennsylvania, and elsewhere are already being used for these purposes. One extraordinary aspect of the Internet is that for the price of a local phone call people can interact with each other not only in their own communities but also with those who have comparable interests throughout the nation and throughout the world. Properly constructed to accommodate civic as well as consumer needs, the information superhighway can become a remarkable force for democratic involvement and action.

In that respect, government can take a leaf from modern-day business practices that increasingly seek to bring customers and employees directly into the process of designing products, planning services, and making their own decisions about priorities. Consumers become part of the product design team when they are consulted about what features they want most in a car, in a computer keyboard, or in a remote control device. Workers on the factory floor are asked to recommend better ways to make the products they turn out on the grounds that they are in the best position to know.

Traditionally, government has completely failed to consult its own constituents when developing new regulations or new policies. Public officials have neglected to bring into the planning process the very people who will be most affected by the actions they take. In the new telecommunications era, that is bound to change. The job of the Federal Commission on Citizenship will be to develop new political systems and recommend interactive communications mechanisms to get the public involved without creating gridlock from the bottom up and incapacitating government's ability to get anything done.

RESHAPING DEMOCRATIC PROCESSES
FOR THE TELECOMMUNICATIONS AGE

The Constitution makes no provision for the use of federal ballot initiatives or national referenda. It was careful to establish the United States as a representative republic, not a direct democracy. Nevertheless, daily public opinion polls—although they have no official standing and are often self-serving, misleading, and inaccurate—play an increasingly important role in influencing the daily decisions of government. And, as we have pointed out, active forms of public expression—E-mail and fax messages to the White House, telephone calls and on-line conversations with congressional offices, for example—also serve to plunge the public at large directly into the decision-making process. However, public opinion polls, along with counting up the number of pro and con spontaneous or orchestrated messages from the electorate, are at best inadequate substitutes for actual voting or other official methods of tabulating the public's views in civic proceedings. Is there a new role to be played at the federal level by officially authorized national plebiscites, referenda, and ballot initiatives in our representative republic? Soon it will be technically feasible to have instant tele-voting from home or workplace, using personal computers and microprocessors. With a system of personal codes and the right telecommunications technology in the hands of all citizens, no longer would there be a practical reason why voting should necessarily be limited to the first Tuesday after the first Monday in November. The people could be asked to vote electronically on any specific measure any time a public response is required. That it is technically feasible, however, does not mean it is politically possible or even desirable.

Should we make provision for popular electronic balloting in our representative system? And if so, on what basis and how often? Should a limit be imposed on how many popular votes can be held in a year? (Congressman Gephardt suggested a maximum of three federal initiatives over several years.) Should the popular votes be advisory only, to help inform the executive branch and the Congress about what course their constituents believe they should follow? Or should certain popular votes be viewed by Congress as mandates or instructions from the people—the new fourth branch of government—subject to being overruled only by presidential veto or the courts? Should the electorate also be given the right to veto unpopular laws that Congress has passed and the president signed—the ultimate popular check in our system of checks and balances? And should the national electorate, through such votes, be able to insist that Congress address legislation that it has decided to ignore or refused to consider (a process that already exists in several states)? All these issues are inevitably going to be raised in the electronic republic and need to be thoroughly explored by the proposed federal commission.

In addition to Ross Perot, a growing number of politicians on both sides of the aisle now advocate the use of federal referenda and initiatives to vote on new taxes, prayer in the schools, and other highly controversial issues. When the ballot initiative process first was introduced in California in the early 20th century, the belief was that it would act only as a safety valve, enabling citizens "to supplement the work of the legislature by initiating those measures which the legislature either viciously or negligently fails or refuses to enact." During the past two decades, however, the California legislature has taken a back seat to the initiative process, which has escalated into "a major if not the principal generator of important state policy, while state government often sits as an understudy responding . . . in a supplemental and reactive fashion." As a result, the legislative process has been put directly into the hands of the people. Referenda have turned into what Republican Jack Kemp calls "the cutting edge of the effort to enact conservative reforms such as school choice and tax limitation." Opponents, on the other hand, now view referenda and ballot initiatives not as populist measures, but mostly as vehicles for private interests with large political war chests.

In England, advocates of greater direct democracy have urged that citizens be given the right to use newly installed interactive cable technology to call referenda in order to veto parliamentary laws and government decisions they do not like. In 1994, these advocates proposed that a petition with one million signatures should be able to trigger a national electronic referendum for a voter veto. Other proposals would limit the voters' judgment to being "purely consultative." Advocates assert that this process "would give the governed more control over the governors, promote civic education and force politicians to see voters as partners rather than as an audience." The *Economist* commented that the introduction of referenda, ballot initiatives, or voter vetoes certainly might "inject some ancient-Greek directness into . . . tired and uninspiring democratic institutions."

As we have seen, the increasing use of direct popular votes to displace legislative and executive powers has its detractors as well as supporters. Problems include the difficulty of framing suitable referendum questions about complex issues; the inadequacy of the public's knowledge and experience involving many of the issues to be decided; concern over encouraging narrowly focused, single-issue politics; fear of one-sided, high spending, distorting campaigns; lack of opportunity for suitable deliberation and compromise under a system in which the only choices are a vote for or against; and the worrisome potential for the exercise of excessive influence by special interests or precipitant action by an unchecked majority.

While most politicians, as would be expected, resist the idea of imposing greater limits on their power through these direct democracy devices, voters when asked consistently have expressed their overwhelming approval. Surveys suggest that a majority in both the United States and Great Britain would welcome greater use of initiatives and referenda to reduce the power of their elected officials, just as the majority in this country consistently favor term limits.

Assuming that public sentiment for more citizen control over government continues to grow, it seems entirely possible, if not likely, that new procedures for direct voter participation in federal decision making will be introduced before long. At the federal level, at least, that would represent a momentous change in our traditional form of representative government. It would be the job of the Federal Commission on Citizenship to address these issues that are bound to deeply affect the democratic process in the electronic republic.

RESTORING CONCERN
FOR THE *COMMON* INTEREST
AND THE *GENERAL* PUBLIC GOOD

A good deal of political ferment is taking place today at the state and local levels. In 1994, the governors of Massachusetts, Michigan, and Wisconsin easily won reelection, partly because voters thought their innovative policies of fiscal restraint and welfare reform were far ahead of the federal government's efforts in these areas. Health plans in Minnesota, Florida, Washington, and Oregon have moved well ahead of the national debate. Several states, including Virginia and Oregon, outpace the nation in developing plans to provide training and employment for the unskilled. California and, to a lesser extent, Texas and Florida are prompting the entire nation to reconsider its immigration policies. One major role of the information superhighway must be to transport news of local and state initiatives and successes across the nation, a role for which the national media have so far proven to be remarkably inept.

As we have suggested, most political constituencies that arise today no longer are defined by physical proximity and physical borders. Most of the dominant electronic media "have no sense of place" and therefore little geographical definition or identity. Even the most influential newspapers, such as the *Wall Street Journal, The New York Times,* and *USA Today* increasingly are national in scope. Citizens who are widely dispersed tend to focus on and be organized around powerful single issues such as entitlement payments to the aged, AIDS, breast cancer research, abortion, health care reform, prayer in the schools, educational choice, the environment, and gun control. These issues and their advocates transcend political borders. Specific national communities of interest arise with little relevance to the general public interest within states, counties, precincts, and election districts that have traditionally defined the American political system. In the 1994 bi-election, the Republican strategy of nationalizing local campaigns by means of its "Contract with America" was given a good deal of the credit for their overwhelming congressional election victory.

We are approaching a society increasingly characterized by single-issue politics, as local proximity and local identity are diminishing in importance. Concern for the common interest and the general welfare is being displaced by a pervasive focus on individual interests that are capable of being mobilized across the nation. In such an environment, the question arises, how do we form communitywide coalitions and make the political compromises and accommodations that are necessary for the *general interest* to be served? How can the federal system adapt to survive in the electronic republic? The Constitution created a "delicate balance" between small states and big states, local governments and the federal government, as well as among the executive, legislative, and judicial branches. That balance, crafted by the Founders to serve the interest of the nation as a whole as well as of the individual states, threatens to become undone in the telecommunications age.

NBC's former head of research used to say, only somewhat facetiously and with a bit of hyperbole to make his point, that since no New Jersey television stations exist, the state of New Jersey no longer exists. Instead, New Jersey has been carved up between two major metropolitan market areas of dominant media influence—New York and Philadelphia. In the world of new telecommunications media markets, traditional political boundary lines lose much of their relevance.

Distance no longer acts as an obstacle to forming political alliances and organizing political leverage. Today, electronic messages sent across the street or across the world are transmitted via satellite and arrive simultaneously, with equal clarity and timeliness. Using the Internet and other on-line services, people interact with others thousands of miles away who share their same interests as easily as they interact with their neighbors next door. The cost disparities between local and long distance communications are fast disappearing. "Location, location, location" means almost everything in real estate, but almost nothing in telecommunications, and therefore, less than it once did in politics. Many candidates running for local and statewide office receive more campaign contributions from outside their election districts than from within.

One consequence of this trend: The balance of power between the states and the federal government, so carefully specified in the Constitution, has been turned upside down. The dominance of national media means that most members of the electorate know far more about the activities of the president, the cabinet, and the Congress in Washington, than about the close-to-home activities of their own precinct leaders, district representatives, county officials, and state governors and legislators. Few states have statewide C-SPANs or news channels of their own. Proponents of the California Channel cite its need because "most Californians see their state government as only an occasional 15-second television news blip sandwiched between the latest murder and most recent fire."

Even the most sophisticated voters find that what goes on nearby in local government and state capitals is less visible and more incomprehensible than what goes on far afield in national politics. Electronics makes the distant seem remarkably clear and familiar, while the local seems strangely opaque and faraway. This is exactly the opposite of what the Federalist founders intended. The multi-state federal nation, whose very size and complexity, Madison argued, would serve as the ultimate barrier against a tyranny of the majority, can be traversed with the speed of light in the telecommunications age.

The early 1980s saw a new movement, the "New Federalism," start to shift government programs away from Washington, D.C., and devolve them back to state and local jurisdictions. In the 1990s that movement picked up speed to such an extent that today it is one of the dominant trends in government. As the Founders intended, the states once again are taking the initiative in dealing with major domestic problems, while the federal government cuts back. The idea is to bring these programs closer to home, where the people who use them can control them, and where presumably they will be more responsive to people's needs. But can a political system that has grown so dependent on national and even global telecommunications, which perversely perform at their worst in transmitting intelligent information about local governance, actually reverse course and "go home" again? Neighborhood and suburban weeklies and local newspapers offer significant local news coverage, of course. But state and local governments urgently need to build new channels of electronic communication to their own constituents. Modern local video links are essential to transmit local activities and initiatives into citizens' homes. The national information infrastructure must be designed with a strong and complex web of local telecommunications arteries that will carry local traffic and support the exercise of effective local citizenship.

As we go about the complicated task of reshaping representative government and redistributing political power in the electronic republic, we must retain the delicate constitutional balance between local and national, between private interests and the public good, and between minority freedom and majority rule. Those will not be easy tasks. But we cannot afford to miss the opportunity to use these new means of communication for the public benefit. We must harness the interactive telecommunications system to help make modern deliberative democracy satisfy the needs of far more citizens than it does today.

SOURCE NOTES

INTRODUCTION

Page
5 "*Extend the sphere*": The Federalist No. 10, *The Federalist Papers by Alexander Hamilton, James Madison, and John Jay* (New York: Bantam Books, 1982), p. 48.

5 "*Although limited*": Donald Kagan, *Pericles of Athens and the Birth of Democracy* (New York: Touchstone, 1991), p. 1.

5 "*Constitutional space*": Harvey C. Mansfield, "Newt, Take Note: Populism Poses Its Own Dangers," *Wall Street Journal*, November 1, 1994.

6 "*The advent of the computer*": Sven Birkerts, *The Gutenberg Elegies* (Boston: Faber and Faber, 1994), p. 15.

6 "*kindle[s] a flame*": The Federalist No. 10, p. 46.

7 "*I know of no safe*": Thomas Jefferson, *The Political Writings of Thomas Jefferson: Representative Selections*, Edward Dumbauld, ed. (New York: Liberal Arts Press, 1955), p. 93.

CHAPTER 1:
TRANSFORMING DEMOCRACY—AN OVERVIEW

12 "*to refine and enlarge*": The Federalist No. 10, *The Federalist Papers by Alexander Hamilton, James Madison, and John Jay* (New York: Bantam Books, 1982), pp. 46–47.

13 "*is to expect the impossible*": Plato, *The Republic* (London: Penguin Books, 1987), p. 28.

13 "*every individual is free*": Ibid., pp. 375, 384.

13 *"the public voice"*: The Federalist No. 10, p. 47.
14 *"the Oval Office"*: Katharine Q. Seelye, "The Lobbyists Are the Loudest in the Healthcare Debate," *The New York Times,* August 16, 1994, p. A12.
15 *"pure democracy"*: Benjamin Barber, *Strong Democracy: Participatory Politics for a New Age* (Berkeley: University of California Press, 1984), p. xxii.
16 *"The Internet and other spurs"*: Peter H. Lewis, "Electronic Tie for Citizens and Seekers of Office," *The New York Times,* November 6, 1994, p. 1.
17 *Boorstin contrasts . . . "segregated" experience*: Quoted in Jeffrey B. Abramson, F. Christopher Arterton, and Gary R. Orren, *The Electronic Commonwealth* (New York: Basic Books, 1988), p. 290.
18 *"Go back to the first principles"*: Charles A. Beard and Mary Beard, *The Rise of American Civilization,* vol. 2 (New York: Macmillan, 1927), p. 555.
20 *"the seasoned wisdom"*: Quoted in Thomas E. Cronin, *Direct Democracy: The Politics of Initiative, Referendum, and Recall* (Cambridge, Mass.: Harvard University Press, 1989), p. 247.
21 *American democracy . . . "democratizing itself further"*: Harvey C. Mansfield, "Newt, Take Note: Populism Poses Its Own Dangers," *Wall Street Journal,* November 1, 1994.
21 *"voters now exercise"*: California Commission on Campaign Financing, *Democracy by Initiative* (Los Angeles: Center for Responsible Government, 1992), p. 2.
21 *In the 1994 election*: Letter to the author from Ancil H. Payne, Seattle, Washington, February 15, 1995.
24 *"the ultimate weapon"*: Russell Baker, "The Final Analysis," *The New York Times,* November 15, 1994, p. A29.
24 *monetary policies*: Walter B. Wriston, *Twilight of Sovereignty* (New York: Charles Scribner's Sons, 1992), p. 9.
25 *"new source of wealth"*: Ibid., p. xii.
25 *"fitting analogy for television"*: Annamarie Pluhan, letter to *The New York Times,* published September 2, 1993, p. A12.
25 *history . . . communications media*: See Marshall McLuhan, *Understanding Media* (New York: New American Library, 1964).
31 *"the decline of a genuine politics"*: Phillip Hansen, *Hannah Arendt* (Stanford: Stanford University Press, 1993), p. 51.
31 *"society's main transforming resource"*: Wriston, *Twilight of Sovereignty,* p. 8.

CHAPTER 2:
THE ROOTS OF THE ELECTRONIC REPUBLIC:
DEMOCRACY'S THIRD TRANSFORMATION

33 *"Although in our time"*: Donald Kagan, *Pericles of Athens and the Birth of Democracy* (New York: Touchstone, 1991), p. 2.
35 *"the only significant link"*: Christian Meier, *The Greek Discovery of Politics,* tr. David McLintock (Cambridge, Mass.: Harvard University Press, 1990).

35 *"[T]he citizen's affiliation"*: Ibid., p. 72.

35 *"everyone, not only"*: Ibid., p. 23.

35 *"To this end"*: Ibid., p. 48.

35 *"were deemed capable"*: Ibid., p. 68.

35 *Greeks . . . assemblies*: Robert A. Dahl, *Democracy and Its Critics* (New Haven: Yale University Press, 1989), p. 13.

36 *"The faculty of speech"*: Hannah Arendt, *Between Past and Future* (New York: Penguin Books, 1993), p. 22.

36 *Plato and Aristotle*: Dahl, *Democracy and Its Critics*, p. 13.

36 *Council of 500*: Meier, *The Greek Discovery of Politics*, p. 147.

37 *"A further requirement"*: Ibid., pp. 71–72.

37 *"the education of Greek citizens"*: Ibid., p. 295.

37 *"just as all citizens"*: Ibid.

37 *"Only among us"*: Ibid., p. 141.

37 *"In Athens"*: Arendt, *Between Past and Future*, p. 19.

39 *"The ruler is the people"*: Samuel H. Beer, *To Make a Nation: The Rediscovery of American Federalism* (Cambridge, Mass.: Harvard University Press, 1993), p. 81.

40 *"the collective rationality"*: Ibid., p. 77.

40 *"either by themselves"*: Dahl, *Democracy and Its Critics*, p. 28.

40 *"The instant a people"*: Benjamin R. Barber, *Strong Democracy: Participatory Politics for a New Age* (Berkeley: University of California Press, 1984), p. 145.

40 *"that absolute despotic power"*: Ibid.

40 *"Parliament is sovereign"*: Anthony Lewis, "Staving Off the Silencers," *The New York Times Magazine*, December 1, 1991, p. 72.

41 *Montesquieu,* representative democracy: Beer, *To Make a Nation*, p. 90.

41 *"Above these inferior"*: Ibid., p. 152.

42 *"virtual representation"*: Bernard Bailyn, *The Ideological Origins of the American Revolution* (Cambridge, Mass.: Harvard University Press, 1967), p. 166.

42 *"the grand discovery"*: Dahl, *Democracy and Its Critics*, p. 29.

43 *"The point of representative"*: George Will, *Restoration* (New York: The Free Press, 1992), p. 122.

43 *"the acquired knowledge"*: John Rawls, *A Theory of Justice* (Cambridge, Mass.: Harvard University Press, 1971), p. 234.

43 *"an unqualified complaisance"*: The Federalist No. 71, *The Federalist Papers by Alexander Hamilton, James Madison, and John Jay* (New York: Bantam Books, 1982), p. 363.

43 *"Almost unanimous"*: Charles A. Beard and Mary R. Beard, *The Rise of American Civilization*, vol. 1 (New York: Macmillan, 1927), p. 316.

44 *"It is the duty"*: The Federalist No. 71, p. 363.

44 *Madison, "reason"*: Beer, *To Make a Nation*, p. 365.

44 *"but by a system of election"*: Ibid., p. 340.

45 *"In this first parliament"*: George McKenna, *American Populism* (New York: G. P. Putnam's Sons, 1974), p. 32.

45 *"The Revolution"*: Gordon S. Wood, *The Radicalism of the American Rev-*

   olution (New York: Vintage Books, 1993; originally published by Alfred
   A. Knopf, 1991), p. 8.
46 *"after two centuries"*: McKenna, *American Populism,* p. xxv.
47 *"received dozens"*: John H. Fund, "We Are All Pundits Now," *Wall Street
   Journal,* November 8, 1994, p. A22.
47 *"PeaceNet, a liberal"*: Ibid.
48 *"Telecommunications"*: Dahl, *Democracy and Its Critics,* p. 339.
49 *"Every cook"*: Arendt, *Between Past and Future,* p. 19.

                              CHAPTER 3:
                    THE RISING FORCE OF PUBLIC OPINION

50 *"A new situation"*: Foreword, *Public Opinion Quarterly,* vol. 1, no. 1,
   January 1937, p. 3.
50 *"Impalpable as the wind"*: Quoted in Lindsay Rogers, *The Pollsters* (New
   York: Alfred A. Knopf, 1949), p. 22.
55 *"protector of the Constitution"*: Editorial, *The New York Times,* May 10,
   1980.
57 *"Since Public Opinion"*: Walter Lippmann, *Public Opinion* (New York:
   The Free Press, 1965), p. 161.
59 *"absolutely devastating"*: *Nova,* PBS television series produced by WGBH,
   Boston, February 18, 1992.
60 *Poll deficiencies:* Daniel Coleman, "Pollsters Enlist Psychologists in Quest
   for Unbiased Results," *The New York Times,* September 7, 1993, p. C1.
60 *"a textbook case"*: Ibid.
60 *Advertising Research Foundation Study:* Leo Bogart, *Polls and the Awareness of Public Opinion* (New Brunswick, N.J.: Transaction Publishers,
   1985), p. xxiii.
61 *"Of 7,802 calls"*: Coleman, "Pollsters Enlist Psychologists."
61 *"comic book statistics"*: Bogart, *Polls and the Awareness of Public Opinion,* p. 20.
61 *"the language of baby talk"*: Lindsay Rogers, *The Pollsters* (New York:
   Alfred A. Knopf, 1949), p. 17.
63 *"Nothing is more dangerous"*: Bogart, *Polls and the Awareness of Public
   Opinion,* p. 47.
63 *"The duty . . . of a patriotic statesman"*: Quoted in Rogers, *The Pollsters,* p. 92.
63 *"If I worried about the poll ratings"*: "Excerpts from Clinton's News Conference in the Rose Garden," *The New York Times,* May 15, 1993, p. A8.
64 *"Your representative owes you"*: Sir Edmund Burke, speech to the
   Electors of Bristol, November 3, 1774. Quoted in Rogers, *The Pollsters,*
   pp. 213–17.
65 *"The voice of the people"*: Quoted in Thomas E. Cronin, *Direct Democracy: The Politics of Initiative, Referendum, and Recall* (Cambridge, Mass.:
   Harvard University Press, 1989), p. 60.
65 *"The general tendency"*: Walter Lippmann, *The Public Philosophy* (New

Brunswick, N.J.: Transaction Publishers, 1989), pp. 45–46. Originally published as *Essays in the Public Policy* (Boston: Little, Brown, 1955).

65   *"That a plurality of the people"*: Ibid., p. 41.
65   *"public opinion as a collective"*: Benjamin I. Page and Robert Y. Shapiro, *The Rational Public: Fifty Years of Trends in Americans' Policy Preferences* (Chicago: University of Chicago Press, 1992), p. 15.
67   *State initiative and referenda studies:* Cronin, *Direct Democracy*, p. 89.
67   *"over the long period"*: Ibid., p. 72.
67   *"voters were able"*: Ibid., pp. 72–73.
67   *"One thing is clear"*: Albert H. Cantril, ed., *Polling on the Issues* (Bethesda, Md.: Seven Locks Press, 1980), p. 170.
68   *Schockley study:* Ibid., p. 109.

CHAPTER 4:
THE AGE OF MEDIA POWER

69   *"It sets the agenda"*: Theodore H. White, *The Making of the President 1972* (New York: Atheneum, 1973), p. 26.
70   *"the new elector"*: See Austin Ranney, *Channels of Power* (New York: Basic Books, 1983).
70   *"This is the age of media power"*: Joan Konner, dean, Columbia University Graduate School of Journalism, speech to incoming class, September 1992.
70   *Newspapers and the State Department:* Jeffrey Abramson, F. Christopher Arterton, and Gary R. Orren, *The Electronic Commonwealth* (New York: Basic Books, 1988), p. 76.
71   *the* Richmond Enquirer: A. James Reichley, *The Life of the Parties* (New York: The Free Press, 1992), p. 89.
71   *Duff Green's* Telegraph: Ibid.
71   *"Everywhere [newspapers] owed"*: Daniel J. Boorstin, *The Americans: The Colonial Experience* (New York: Vintage Books, 1964), pp. 324–25.
71   *Newspaper circulation in 1800:* Abramson, Arterton, and Orren, *The Electronic Commonwealth*, p. 75.
71   *Newspaper circulation in 1830:* Michael Schudson, *Discovering the News* (New York: Basic Books, 1978), p. 13.
71   *"when the visor"*: Abramson, Arterton, and Orren, *The Electronic Commonwealth*, p. 75.
72   *"Much was made"*: Kiku Adato, *Picture Perfect* (New York: Basic Books, 1993), p. 71.
72   *Newspaper circulations:* Schudson, *Discovering the News*, p. 13.
72   *The United States became:* Abramson, Arterton, and Orren, *The Electronic Commonwealth*, p. 80.
73   *"sensational," "cheap"*: Schudson, *Discovering the News*, p. 22.
73   *"papers with widely different political allegiances"*: Ibid., p. 4.
73   *"The wire services demanded"*: James W. Carey, "The Press and Public Discourse," *Kettering Review*, Winter 1992, pp. 18–19.
74   *"the changeover"*: Schudson, *Discovering the News*, p. 57.
74   *"money had new power"*: Ibid., p. 43.

74   *"Everything is quiet"*: Ibid., p. 62.

76   *"an unprecedented absence"*: Abramson, Arterton, and Orren, *The Electronic Commonwealth*, p. 84.

79   *"an agent of social change"*: Robert S. Lichter, Linda S. Lichter, and Stanley Rothman, *Watching America* (New York: Prentice-Hall Press, 1991), p. 4.

79   *Reporters' liberal bias*: See Herbert J. Gans, *Deciding What's News* (New York: Vintage Books, 1980); and Edward J. Epstein, *News from Nowhere: Television and the News* (New York: Random House, 1973).

80   *"In 1958, 24 percent"*: Schudson, *Discovering the News*, p. 178.

80   *"Journalists . . . covering national politics"*: Ibid., p. 180.

82   *"The men and women who control"*: Quoted in ibid., p. 184.

82   *"Given the striking similarity"*: Ben H. Bagdikian, *The Media Monopoly*, 3d ed. (Boston: Beacon Press, 1993), p. 4.

85   *"The liberties protected"*: John Rawls, *A Theory of Justice* (Cambridge, Mass.: Harvard University Press, 1971), pp. 225–26.

88   *"the tradition of objectivity"*: Ibid., p. 185.

88   *"The Los Angeles Times"*: Tom Rosenstiel, *Strange Bedfellows: How Television and the Presidential Candidates Changed American Politics, 1992* (New York: Hyperion, 1993), pp. 164–65.

89   *"situational bias"*: Thomas E. Patterson, "Let the Press Be the Press," in *1-800-PRESIDENT: The Report of the Twentieth Century Fund Task Force on Television and the Campaign of 1992* (New York: The Twentieth Century Fund Press, 1993), p. 92.

89   *"The press is like the beam"*: Walter Lippmann, *Public Opinion* (New York: The Free Press, 1922), p. 229.

90   *"Real news is bad news"*: Marshall McLuhan, *Understanding Media* (New York: New American Library, 1964), p. 183.

90   *"the press treats"*: Patterson, "Let the Press Be the Press," p. 97.

92   *"The highest power"*: Epstein, *News from Nowhere*, p. 39.

92   *"Television can give"*: Bernice Buresh, "Stop the World! I Want to Think!," *Los Angeles Times*, December 30, 1989, p. B6.

CHAPTER 5:
TELEVISION AND BEYOND

93   *"the mean levels of concentration"*: Robert Kubey and Mihaly Csikszentmihelyi, *Television and the Quality of Life* (Hillsdale, N.J.: Lawrence Earlbaum Associates, 1990), p. 81.

94   *"The world we have to deal with"*: Walter Lippmann, *Public Opinion* (New York: The Free Press, 1922), p. 18.

95   *"At one time"*: Joshua Meyrowitz, *No Sense of Place* (New York: Oxford University Press, 1985), p. vii.

97   *Newsfilm v. videotape*: Sig Mickelson, *The Electric Mirror* (New York: Dodd, Mead, 1972), p. 54, and *From Whistle Stop to Sound Bite* (New York: Praeger, 1989), p. 104.

97   *CBS's imperious chairman*: Sally Bedell Smith, *In All His Glory* (New York: Simon & Schuster, 1990), p. 185.

100 *"more certain than ever"*: Stanley Karnow, *Vietnam: A History* (New York: Viking, 1983), p. 547.

101 *"electronic tools of the trade"*: Jerry Landay, *Chicago Tribune*, February 1, 1991, p. 15.

102 *smart bombs:* H. Jachim Maitre, "Journalistic Incompetence," *Nieman Reports,* Summer 1991, p. 10.

102 *"What we decided"*: Phillip M. Gollner, "Batons in the Jury Room: Weighing 4 Officers' Fate," *The New York Times,* April 24, 1993, p. A1.

102 *"the prime controllers"*: Herbert J. Gans, "Bystanders as Opinion Makers: A Bottoms-Up Perspective," *Gannett Center Journal,* Spring 1989, p. 97.

105 *News sources:* Michael Schudson, *Discovering the News* (New York: Basic Books, 1978), p. 182.

106 *"wiggle room"*: Robert Michel, "Politics in the Age of Television," *Washington Post National Weekly Edition,* June 4, 1984, p. 27.

106 *"of their most valuable asset"*: William Safire, "The Newt Deal," *The New York Times,* November 5, 1994, p. A27.

108 *"By priming"*: Shanto Iyengar and Donald R. Kinder, *News That Matters* (Chicago: University of Chicago Press, 1987), p. 4.

108 *"The medium creates"*: Robert S. Lichter, Linda S. Lichter, and Stanley Rothman, *Watching America* (New York: Prentice-Hall Press, 1991), p. 4.

108 *"eagerness to tackle"*: Ibid., p. 297.

111 *"The most remote"*: Donald Horton and R. Richard Wohl, quoted in Joshua Meyrowitz, *No Sense of Place,* p. 119.

111 *"great leader"*: Meyrowitz, *No Sense of Place,* p. 269.

112 *"is a shield"*: Ibid., p. 99.

113 *Carter TV talk:* George E. Reedy, "A Symbol Is Worth More Than a 1,000 Words," *TV Guide,* December 31, 1977, p. 2.

114 *"eloquent and lofty manner"*: Alexis de Toqueville, *Democracy in America,* vol. 1 (New York: Alfred A. Knopf, 1945), p. 187.

115 *"mostly an extension"*: Howard Rosenberg, "The Cult of Personality," *American Journalism Review,* September 1993, p. 18.

115 *Local TV news:* Steven Hess, "Television and the Loss of Place," SILHA Lecture, SILHA Center for the Study of Media Ethics and Law, University of Minnesota, 1990, p. 2.

116 *"Newsrooms often are overstocked"*: Ibid., p. 14.

118 *"At one level"*: Max Frankel, "Word and Image," *New York Times Magazine,* November 13, 1994, sec. 6, p. 36.

119 *"dwarfs anything"*: Jeffrey B. Abramson, F. Christopher Arterton, and Gary R. Orren, *The Electronic Commonwealth* (New York: Basic Books, 1988), p. 7.

119 *"the most sweeping test yet"*: John Lippman, "Tuning Out the TV of Tomorrow," *Los Angeles Times,* August 31, 1993, p. A1.

119 *"the tendency is"*: Marshall McLuhan, *Understanding Media* (New York: New American Library, 1964), p. 182.

CHAPTER 6:
DECLINE OF THE OLD POLITICS, THE RISE OF THE NEW

120 *Party membership:* William Greider, *Who Will Tell the People* (New York: Simon & Schuster, 1992), p. 249.

121 *Independent voters:* A. James Reichley, *The Life of the Parties* (New York: The Free Press, 1992), p. 7.

121 *Democratic chairman's complaint:* David Wilhelm, quoted in Jon Meacham, "Why the Party of the People Has a Grassroots Problem," *Washington Monthly,* October 1993, p. 24.

122 *"labor intensive work":* Arthur M. Schlesinger, Jr., *The Cycles of American History* (Boston: Houghton Mifflin, 1986), p. 272.

124 *"organized interest groups":* Reichley, *The Life of the Parties,* p. 7.

124 *Abortion and other issues:* Alan I. Abramowitz, "It's Abortion, Stupid." Paper presented at the 1993 meeting of the American Political Science Association in Washington, D.C.

124 *"are likely to be dominated":* David S. Broder, "Lasting Effects of Perot, Religious Right Debated," *Washington Post,* September 6, 1993, p. A6.

124 *"politics of principle":* V. O. Key, quoted in David S. Broder, *The Party's Over* (New York: Harper & Row, 1971), p. 172.

125 *"The system was simply overwhelmed":* Robin Toner, "Making Sausage: The Art of Reprocessing the Democratic Process," *The New York Times,* September 4, 1994, sec. 4, p. 1.

127 For television's early convention coverage, see Reuven Frank, *Out of Thin Air* (New York: Simon & Schuster, 1991), pp. 1–27.

128 *"miracle" election:* David McCullough, *Truman* (New York: Simon & Schuster, 1992), p. 710.

130 *"Tans, light blues and grays":* Quoted in "Blue Conventions," CBS Television brochure, 1956.

131 *"TV's impact":* Ibid.

135 *"such an overarching event":* Reuven Frank, *Out of Thin Air,* p. 185.

136 *"I have just been informed":* Quoted in Gene Shalit and Lawrence K. Grossman, *Somehow It Works* (Garden City, N.Y.: Doubleday, 1965), Foreword.

138 *"It would not have happened":* Frank, *Out of Thin Air,* p. 272.

139 *"two acceptable methods":* Thomas E. Patterson, *Out of Order* (New York: Alfred A. Knopf, 1993), pp. 31–33.

140 *"the circus":* Gary R. Orren and Nelson W. Polsby, eds., *Media and Momentum: The New Hampshire Primary and Nomination Politics* (Chatham, N.J.: Chatham House, 1987), p. 128.

141 *"a new form":* California Commission on Campaign Financing, *Democracy by Initiative* (Los Angeles: Center for Responsive Government, 1992), p. 1.

CHAPTER 7:
THE SHAPE OF THE ELECTRONIC REPUBLIC:
THE CITIZENS, THE CONGRESS, THE PRESIDENCY,
AND THE JUDICIARY

145 *Home schooling amendment:* "The Electronic Word Went Out," *The New York Times,* March 21, 1994, p. A16.

146 *"The advantages offered": The Promise and Perils of Emerging Information Technologies,* Aspen Institute Communications and Society Program. Forum report by David Bollier, rapporteur, Charles Firestone, director, Washington, D.C., 1993, p. 26.

147 *"The country may be moving":* Ted Koppel, "The Perils of Info-Democracy," *The New York Times,* July 1, 1994, p. A25.

149 *"A person who taps":* Edmund L. Andrews, "Mr. Smith Goes to Cyberspace," *The New York Times,* January 6, 1995, p. A22.

150 *"eliminated or superseded": Pacific States Telephone and Telegraph Co.* v. *Oregon,* as quoted in California Commission on Campaign Financing, *Democracy by Initiative* (Los Angeles: Center for Responsive Government, 1992), pp. 301–2.

151 *"country needs to make":* Christopher Georges, "Perot and Con," *Washington Monthly,* June 1993, p. 39.

151 *"The ballot initiative":* California Commission on Campaign Financing, *Democracy by Initiative,* p. 1.

152 *Ludlow Amendment:* Thomas E. Cronin, *Direct Democracy: The Politics of Initiative, Referendum, and Recall* (Cambridge, Mass.: Harvard University Press, 1989), pp. 169–70.

152 *Gephardt plan:* Ibid., p. 172.

153 *Nova Scotia Liberal Party:* Lloyd N. Morrisett, president's essay, John and Mary R. Markle Foundation Annual Report, 1991–92, p. 5.

155 *"Old citizens":* John Boardman, Jasper Griffin, and Murray Oswyn, eds., *The Oxford History of Greece and the Hellenistic World* (New York: Oxford University Press, 1991), p. 194.

155 *"doubling or tripling":* "Want a Political Revolt? Here's a Place to Start," editorial, *USA Today,* November 8, 1994, p. 10A.

155 *"today's technology":* Speech by Vice President Al Gore at the Academy of Television Arts and Sciences, Los Angeles, January 11, 1994.

156 *"In the worst case":* E. J. Dionne, columnist for the *Washington Post,* as quoted in the *Annenberg Washington Program Annual Report, 1994,* Annenberg Washington Program, Communications Policy Studies, Washington, D.C., 1994.

158 *Lamar Alexander meetings:* John H. Fund, "We Are All Pundits Now," *Wall Street Journal,* November 8, 1994, p. A22.

158 *"The present system":* James Rogers, "Congress Go Home," *Sacramento Union,* July 29, 1988, p. A13.

159 *"Neither the President":* *Nieman Reports,* Winter 1993, p. 57.

159 *"Clinton's team":* Mark Miller, "And You Thought the Campaign Was Long," *Newsweek,* January 18, 1993, p. 25.

159 *"a war room":* Paul Brace and Barbara Hinckley, *Follow the Leader: Opin-*

ion Polls and the Modern Presidents (New York: Basic Books, 1992), p. vi.

159 *"Behind all foreign policy"*: R. W. Apple, Jr., "Clinton Looks Homeward," *The New York Times,* January 13, 1994, p. A1.

159 *"If you want to change"*: Sidney Blumenthal, "The Education of a President," *The New Yorker,* January 24, 1994, p. 40.

159 *looked back with longing*: William Safire, "Clinton's Voodoo Intervention," *The New York Times,* September 12, 1994, p. A15.

160 *"plebiscitary" presidency*: James S. Fishkin, *Democracy and Deliberation: New Directions for Democratic Reform* (New Haven: Yale University Press, 1991), pp. 46, 48–49.

160 *"the kind of leader"*: Benjamin R. Barber, *Strong Democracy: Participatory Politics for a New Age* (Berkeley: University of California Press, 1984), p. 239.

160 *"facilitating leadership"*: Meg Greenfield, "Talk Government," *Newsweek,* March 15, 1993, p. 84.

161 *"moment by moment"*: Brace and Hinckley, *Follow the Leader,* p. 3.

161 *"mainly a follower"*: Gary Wills, *Certain Trumpets* (New York: Simon & Schuster, 1994), p. 12.

161 *"mobilize others"*: Ibid., p. 17.

161 *"a mutually determinative"*: Ibid., p. 14.

161 *"the landscape of American politics"*: *Nieman Reports,* Winter 1993, p. 57.

162 *"hired help"*: Mailer from Project Vote Smart, Portland, Oregon, August 24, 1993.

163 *"a nation that traces power"*: Julian Eule, "Judicial Review of Direct Democracy," *99 Yale Law Journal,* issue 7, May 1989, pp. 1504, 1506.

163 *"While the initiative"*: Donald S. Goldberg, as quoted in California Commission on Campaign Financing, *Democracy by Initiative,* pp. 304–5.

163 *Role of courts*: See ibid.

CHAPTER 8:
THE PERILS AND PROMISE
OF THE ELECTRONIC REPUBLIC

165 *"an Athens without slaves"*: Howard Rheingold, *The Virtual Community: Finding Connection in a Computerized World* (Reading, Mass.: Addison-Wesley, 1993), p. 279.

165 *"The push of the new technologies"*: W. Russell Neuman, *The Future of the Mass Audience* (Cambridge, England: Cambridge University Press, 1993), p. 13.

165 *"a major force for freedom"*: George Gilder, *Life After Television* (Knoxville, Tenn.: Whittle Direct Books, 1990), p. 26.

166 *"intellectual leverage"*: Rheingold, *The Virtual Community,* p. 4.

166 *"an America where poor"*: Editorial, "Mr. Gore's Video Vision," *The New York Times,* January 17, 1994, p. A16.

166 *"a great flowering"*: *Economist,* January 12, 1994, p. 3, special sec.

166 *"the telephone, the telegraph"*: James A. Monroe, *The Democratic Wish:*

Popular Participation and the Limits of American Government (New York: Basic Books, 1992), p. 109.

166 *"It is inconceivable"*: Quoted in Leo Bogart, *Commercial Culture* (New York: Oxford University Press, 1995), p. 69.

167 *"the great radiance in the sky"*: Edward R. Murrow speech to Radio and Television News Directors Association, Chicago, October 15, 1958.

167 *"The place to sit"*: Lawrence K. Grossman, "The Best Seat in the House?," *Media Studies Journal*, Winter 1995, p. 49.

169 *"vast wasteland"*: Newton H. Minow speech to National Association of Broadcasters, New York, May 9, 1961.

169 *"Commercial television"*: Walter Goodman, "What's Bad for Politics Is Great for Television," *The New York Times*, November 27, 1994, Arts and Leisure sec., p. 33.

169 *"These are not novel"*: Ibid.

170 *"The tendency is definitely"*: Monroe E. Price, "Free Expression and Digital Dreams: The Open and Closed Terrain of Speech," New York, Cardozo Law School, unpublished paper, p. 4.

171 *"The present quality of television"*: Leo Bogart, "Highway to the Stars or Road to Nowhere?" *Media Studies Journal*, Winter 1994, p. 6.

171 *"misled by the artful"*: Federalist No. 63, *The Federalist Papers by Alexander Hamilton, James Madison, and John Jay* (New York: Bantam Books, 1982), p. 320.

171 *"the degrading effects"*: Bogart, *Commercial Culture*, p. 9.

172 *"fleeting, disjointed, visual glimpses"*: George F. Kennan, "If TV Drives Foreign Policy, We're in Trouble," *The New York Times*, October 1993, sec. 4, p. 14.

172 *"democracy"* . . . *"low connotations"*: Charles A. Beard and Mary R. Beard, *America in Mid-Passage* (New York: Macmillan, 1939), pp. 922–24.

173 *"The government"*: Ibid., p. 932.

173 *"two or three companies"*: Doug Halonen, "Malone: Few Will Rule Superhighway," *Electronic Media*, January 31, 1994, p. 31.

174 *"Monopoly is a terrible thing"*: Editorial, "Sweet Are the Uses of Monopoly," *Index on Censorship*, April–May 1994, p. 3.

174 *Wright testimony*: Federal Communications Commission, *FCC 1990 Cable Report*, cited in para. 120, fn. 85.

174 *"would have liked"*: Joe Flint, "Summit Sees Bright Future for Info Highway," *Broadcasting and Cable Magazine*, January 17, 1994, p. 8.

174 *"There are at least four companies"*: Don West and Joe Flint, "The Wonder's Still Wireless," *Broadcasting and Cable Magazine*, January 24, 1994, p. 22.

175 *"The cable model"*: Henry Geller, *1995–2005: Regulatory Reform for Principal Electronic Media* (Washington, D.C.: The Annenberg Washington Program in Communications Policy Studies of Northwestern University, 1994), p. 34.

179 *"the telefuture"*: Gilder, *Life After Television*, p. 24.

180 *"To the Union Pacific"*: Charles A. Beard and Mary R. Beard, *The Rise of American Civilization*, vol. 2 (New York: Macmillan, 1927), p. 128.

183 *"political information"*: Benjamin I. Page and Robert Y. Shapiro, *The Ra-*

tional Public: Fifty Years of Trends in Americans' Policy Preferences (Chicago: University of Chicago Press, 1992), p. 395.

184 *"replaced political parties"*: Russell Baker, "The Flexible Goodbye," *The New York Times,* July 26, 1994, p. A19.

185 *"Money often dominates"*: California Commission on Campaign Financing, *Democracy by Initiative* (Los Angeles: Center for Responsive Government, 1992), pp. 263–64.

185 *State initiatives study:* Christopher Georges, "Perot and Con," *Washington Monthly,* June 1993, pp. 41–42.

185 *Overturning campaign finance reform:* Buckley v Valeo, 424, US 1 (1976); and *National Bank v Bellotti,* 433 US 765 (1978).

186 *"cannot help but exacerbate"*: Oscar H. Gandy, Jr., *The Panoptic Sort: A Political Economy of Personal Information.* (Boulder, Colo.: Westview Press, 1993), pp. 1–2.

186 *"may distort"*: Page and Shapiro, *The Rational Public,* p. 394.

186 *"On the one hand"*: Ibid., pp. 381–82.

187 *"the empty politics"*: William Greider, *Who Will Tell the People* (New York: Simon & Schuster, 1992), p. 264.

188 *"Immediate reward news"*: Jurgen Habermas, *The Structural Transformation of the Public Sphere: An Inquiry into a Category of Bourgeois Society,* tr. Thomas Burger (Cambridge, Mass.: MIT Press, 1993), p. 169.

189 *Television survey: Times Mirror Media Monitor,* March 24, 1993.

CHAPTER 9:
MEDIA REFORM—BACK TO THE FUTURE

191 *"to order, access, store"*: Robert Pepper, "Broadcasting Policies in a Multichannel Marketplace," in Charles M. Firestone, ed., *Television for the 21st Century: The Next Wave* (Washington D.C.: The Aspen Institute Communications and Society Program, 1993), p. 120.

192 *"the linear descendants"*: Lucas A. Powe, Jr., *American Broadcasting and the First Amendment* (Berkeley: University of California Press, 1987), pp. 29–30.

193 *"a law unto itself"*: Kovacs v. Cooper, 336 US 77, 97 (1949), Jackson J. concurring.

193 *"differences in the First Amendment"*: Red Lion Broadcasting Co., 395 US 1 at 386.

193 *"a separate First Amendment"*: Robert Corn-Revere, "New Technology and the First Amendment—Breaking the Cycle of Repression," Columbia Institute for Tele-Information Conference on the 1992 Cable TV Act, Columbia University, February 25, 1994, p. 43.

193 *"has led to a scholastic set"*: Ithiel de Sola Pool, *Technologies of Freedom* (Cambridge, Mass.: Harvard University Press, 1983), p. 250.

193 *"reveal curious judicial blindness"*: Ibid., p. 19.

194 *"If I were to pose"*: Henry Geller, *1995–2005: Regulatory Reform for Principal Electronic Media"* (Washington, D.C.: The Annenberg Washington

Program in Communications Policy Studies of Northwestern University, 1991), p. 13.

194 *a charade:* Ibid., p. 15.

194 *"the privilege to broadcast":* Powe, *American Broadcasting and the First Amendment,* p. 6.

197 *"As adamant as my country has been":* William J. Brennan, Jr., Freedom of Speech Symposium, Hebrew University, Jerusalem, December 23, 1987.

200 *media ownership:* Ben H. Bagdikian, *The Media Monopoly,* 3d ed. (Boston, Beacon Press, 1993), pp. 3–4, 18.

204 *"the only organized private business":* Potter Stewart, "Or of the Press," 26 *Hastings Law Journal* 633, 1975.

205 *"The FCC, the SEC":* Letter to the author from Ancil H. Payne, Seattle, Washington, February 15, 1995.

205 *"I'll tell you":* Edmund L. Andrews, "Mr. Smith Goes to Cyberspace," *The New York Times,* January 6, 1995, p. A22.

206 *Moore's Law:* Geller, *1995–2005: Regulatory Reform for Principal Electronic Media,* p. 7.

207 *"Congress could reasonably":* Ibid., p. 21.

209 *"educational and informational":* Edmund L. Andrews, "Broadcasters, to Satisfy Law, Define Cartoons as Educational," *The New York Times,* September 30, 1992, p. A1.

209 *"Commercial broadcasters should welcome":* Geller, *1995–2005: Regulatory Reform for Principal Electronic Media,* p. 23.

209 *"the time has come":* "Hundt's New Deal," *Broadcasting and Cable Magazine,* August 1, 1994, p. 6.

210 *if the compact is broken: Communications Daily,* October 20, 1994, p. 9.

210 *"But I mean":* Thomas Jefferson, Letter to Colonel Edward Carrington, January 16, 1787. Quoted in George McKenna, *American Populism* (New York: G. P. Putnam's Sons, 1974), p. 22.

211 *"Public Television programming":* Carnegie Commission on Educational Television, *Public Television, A Program for Action* (New York: Bantam Books, 1967), p. 92.

213 *"the broad national values":* The Twentieth Century Fund Task Force on Public Television, *Quality Time?* (New York: The Twentieth Century Fund Press, 1993), p. 3.

214 *"increase voter participation":* Flier describing the Democracy Channel, Center for Governmental Studies, Los Angeles, 1994.

216 *Federal funding of local public television stations:* The Twentieth Century Fund Task Force, *Quality Time?,* p. 49.

217 *democracy "is a value":* Corn-Revere, "New Technology and the First Amendment," p. 97.

217 *"there is an important role":* Ibid., p. 98.

CHAPTER 10:
NONGOVERNMENTAL AND OTHER REFORMS

220　*"suggested I take"*: Letter to the author from Ancil H. Payne, Seattle, Washington, February 15, 1995.

220　*"No one was satisfied"*: Tim Russert, "For '92, the Networks Have to Do Better," *The New York Times*, March 4, 1990.

221　*Libel suits*: Quoted in Barbara Dill, "Libel Law Doesn't Work, but Can It be Fixed?," in *At What Price? Libel Law and Freedom of the Press* (New York: The Twentieth Century Fund Press, 1993), p. 69.

223　*"malice"*: Ibid.

235　*Initiative campaign financing*: California Commission on Campaign Financing, *Democracy by Initiative* (Los Angeles: Center for Responsive Government, 1992), p. 296.

235　*Finance reforms*: Anthony Corrado, *Paying for Presidents* (New York: The Twentieth Century Fund Press, 1993), p. 1.

236　*"the corrosive and distorting effects"*: *Austin v Michigan Chamber of Commerce*, 110 S. Ct. 1391 and 1398 (1990).

CHAPTER 11:
"TO INFORM THEIR DISCRETION"

237　*"[T]he street, the city"*: Stephen Spender, *World Within World* (New York: St. Martin's Press, 1994), p. xiii.

238　*"The cumulative effect"*: Michael Wines, "Washington Really Is in Touch: We're the Problem," *The New York Times*, October 16, 1994, sec. 4, p. 1.

238　*"How much of the time"*: "The Anger: Ever Deeper," *New York Times Magazine*, October 16, 1994, p. 37.

239　*"in splendid isolation"*: Wines, "Washington Really Is in Touch."

239　*"perhaps the only means"*: Alexis de Tocqueville, quoted in David S. Broder, *The Party's Over* (New York: Harper & Row, 1971), p. 263.

240　*"To take a hand"*: Alexis de Tocqueville, *Democracy in America*, vol. 1 (New York: Alfred A. Knopf, 1945), pp. 250–51.

240　*"inquiry and communication"*: John Dewey, quoted in Benjamin I. Page and Robert Y. Shapiro, *The Rational Public: Fifty Years of Trends in Americans' Policy Preferences* (Chicago: University of Chicago Press, 1992), p. 363.

240　*"the council state"*: Hannah Arendt, *Crises of the Republic* (San Diego: Harcourt Brace Jovanovich, 1972), pp. 232–33. And see Phillip Hansen, *Hannah Arendt* (Stanford: Stanford University Press, 1993).

241　*"the deliberative poll"*: James S. Fishkin, *Democracy and Deliberation: New Directions for Democratic Reform* (New Haven: Yale University Press, 1991), p. 3.

241　*"minipopulus"*: Robert A. Dahl, *Democracy and Its Critics* (New Haven: Yale University Press, 1989), pp. 340–41.

244　*"In Plato's Republic"*: Ibid., p. 53.

244 *"not as someone"*: Donald Kagan, *Pericles of Athens and the Birth of Democracy* (New York: Touchstone, 1991), p. 144.

247 *"The strength and spring"*: Moses Mather, quoted in Gordon S. Wood, *The Creation of the American Republic* (New York: W. W. Norton, 1969), p. 120.

250 *"to supplement the work"*: California Commission on Campaign Financing, *Democracy by Initiative,* quoting California special election ballot pamphlet "Arguments in Favor of SCA 22, October 11, 1911," p. 53.

250 *"a major if not the principal"*: Ibid., p. 53.

250 *"the cutting edge"*: Jack Kemp, "GOP Contract: My Amendments," *Wall Street Journal*, September 23, 1994.

250 *"purely consultative"*: *Economist*, September 17, 1994, p. 63.

250 *"would give the governed"*: Ibid., p. 64.

250 *"inject some ancient-Greek"*: Ibid., p. 63.

253 *"most Californians"*: Description of the California Channel, Center for Responsive Government, 1989, p. 2.

BIBLIOGRAPHY

Abel, Elie, ed. *What's News*. San Francisco: Institute for Contemporary Studies, 1981.

Abramson, Jeffrey B., F. Christopher Arterton, and Gary R. Orren. *The Electronic Commonwealth*. New York: Basic Books, 1988.

Adams, William C., ed. *Television Coverage of the 1980 Presidential Campaign*. Norwood, N.J.: Ablex Publishing, 1983.

Adato, Kiku. *Picture Perfect*. New York: Basic Books, 1993.

Alexander, Yonah, and Robert G. Picard. *In the Camera's Eye*. Washington, D.C.: Brassey's (U.S.), 1991.

Alger, Dean. *The Media, the Public and the Development of Candidates' Images in the 1992 Presidential Election*. Cambridge, Mass.: Joan Shorenstein Barone Center on Press, Politics and Public Policy, Harvard University, 1994.

Alinsky, Saul D. *Rules for Radicals*. New York: Vintage Books, 1989.

Alterman, Eric. *Sound and Fury*. New York: HarperCollins, 1992.

Altschull, Herbert J. *From Milton to McLuhan*. New York: Longman, 1990.

Araskog, Rand V. *The ITT Wars*. New York: Henry Holt, 1989.

Arendt, Hannah. *Between Past and Future*. New York: Penguin Books, 1993.

———. *Crises of the Republic*. San Diego: Harcourt Brace Jovanovich, 1972.

Arterton, F. Christopher. *Teledemocracy*. Newbury Park, Calif.: Sage Publications, 1987.

Bachrach, Peter. *The Theory of Democratic Elitism*. Boston: Little, Brown, 1967.

Bagdikian, Ben H. *The Information Machines*. New York: Harper & Row, 1971.

———. *The Media Monopoly*, 3d ed. Boston: Beacon Press, 1993.

Bailyn, Bernard. *The Ideological Origins of the American Revolution*. Cambridge, Mass.: Harvard University Press, 1967.

Barber, Benjamin R. *Strong Democracy: Participatory Politics for a New Age.* Berkeley: University of California Press, 1984.

Barber, James David. *The Presidential Character.* Englewood Cliffs, N.J.: Prentice-Hall, 1985.

Barnouw, Erik. *The Sponsor.* New York: Oxford University Press, 1978.

———. *Tube of Plenty.* New York: Oxford University Press, 1975.

Beard, Charles A., and Mary R. Beard. *America in Mid-Passage.* New York: Macmillan, 1939.

———. *The Beards' Basic History of the United States.* Garden City, N.Y.: Doubleday, 1944.

———. *The Rise of American Civilization.* New York: Macmillan, 1927.

Bedell, Sally. *In All His Glory: The Life of William Paley.* New York: Simon & Schuster, 1990.

———. *Up the Tube.* New York: Viking, 1981.

Beer, Samuel H. *To Make a Nation: The Rediscovery of American Federalism.* Cambridge, Mass.: Harvard University Press, 1993.

Bell, Jeffrey. *Populism and Elitism.* Washington, D.C.: Regnery Gateway, 1992.

Benjamin, Burton. *Fair Play.* New York: Harper & Row, 1988.

Bennett, W. Lance. *News: The Politics of Illusion.* New York: Longman, 1988.

Bergreen, Laurence. *Look Now, Pay Later.* Garden City, N.Y.: Doubleday, 1980.

Berns, Walter. *The First Amendment and the Future of American Democracy.* New York: Basic Books, 1976.

Besen, Stanley M., et al. *Misregulating Television.* Chicago: University of Chicago Press, 1984.

Birkerts, Sven. *The Gutenberg Elegies.* Boston: Faber and Faber, 1994.

Blumenthal, Sidney. *Our Long National Daydream.* New York: HarperCollins, 1990.

Boardman, John, Jasper Griffin, and Oswyn Murray, eds. *The Oxford History of Greece and the Hellenistic World.* Oxford, England: Oxford University Press, 1991.

Bogart, Leo. *Commercial Culture.* New York: Oxford University Press, 1995.

———. *Polls and the Awareness of Public Opinion.* New Brunswick, N.J.: Transaction Publishers, 1988.

Bollier, David, and Charles M. Firestone. *The Promise and Perils of Emerging Information Technologies.* Washington, D.C.: The Aspen Institute, 1993.

Boorstin, Daniel J. *The Americans: The Colonial Experience.* New York: Vintage Books, 1964.

———. *The Image: A Guide to Pseudo-Events in America.* New York: Vintage Books, 1992.

Bower, Robert T. *Television and the Public.* New York: Holt, Rinehart & Winston, 1973.

Brace, Paul, and Barbara Hinckley. *Follow the Leader: Opinion Polls and the Modern Presidents.* New York: Basic Books, 1992.

Bradlee, Benjamin C. *The Theodore H. White Lecture,* Cambridge, Mass.: Joan Shorenstein Barone Center on the Press, Politics and Public Policy, Harvard University, 1991.

Braestrup, Peter. *Battle Lines: Report of the Twentieth Century Fund Task Force on the Military and the Media.* New York: Priority Press, 1985.

Brand, Stewart. *The Media Lab*. New York: Viking Penguin, 1987.

Broder, David S. *Behind the Front Page*. New York: Simon & Schuster, 1987.

———. *The Party's Over*. New York: Harper & Row, 1971.

Brown, Lester L. *Les Brown's Encyclopedia of Television*. Detroit: Visible Ink Press, 1992.

———. *Television: The Business Behind the Box*. New York: Harcourt Brace Jovanovich, 1971.

Buchanan, Bruce. *Electing a President: The Markle Commission Research on Campaign '88*. Austin: University of Texas Press, 1991.

Burns, James MacGregor. *The Deadlock of Democracy*. Englewood Cliffs, N.J.: Prentice-Hall, 1963.

California Commission on Campaign Financing. *Democracy by Initiative*. Los Angeles: Center for Responsive Government, 1992.

Cantril, Albert H., ed. *Polling on the Issues*. Bethesda, Md.: Seven Locks Press, 1980.

Cantril, Hadley. *Gauging Public Opinion*. Princeton: Princeton University Press, 1944.

Carnegie Commission on Educational Television. *Public Television: A Program for Action*. New York: Bantam Books, 1967.

Carpenter, Edmund, and Marshall McLuhan, eds. *Explorations in Communication*. Boston: Beacon Press, 1960.

Chafee, Zechariah, Jr. *Freedom of Speech*. New York: Harcourt, Brace & Howe, 1920.

Chester, Lewis, Godfrey Hodgson, and Bruce Page. *An American Melodrama*. New York: Viking, 1969.

Chomsky, Noam. *Necessary Illusions*. Boston: South End Press, 1989.

Cirino, Robert. *Don't Blame the People*. Los Angeles: Diversity Press, 1971.

Clurman, Richard M. *Beyond Malice*. New Brunswick, N.J.: Transaction Publishers, 1988.

Coates, Vary T., and Bernard Finn. *A Retrospective Technology Assessment: Submarine Telegraphy*. San Francisco: San Francisco Press, 1979.

Cohen, Jeremy. *Congress Shall Make No Law*. Ames: Iowa State University Press, 1989.

Commager, Henry Steele. *Commager on Tocqueville*. Columbia: University of Missouri Press, 1993.

Cook, Philip S., Douglas Gomery, and Lawrence W. Lichty. *The Future of News*. Washington, D.C.: The Woodrow Wilson Center Press, 1992.

Cornford, Francis MacDonald, trans. *The Republic of Plato*. London: Oxford University Press, 1945.

Corrado, Anthony. *Paying for Presidents*. New York: The Twentieth Century Fund Press, 1993.

Cronin, Thomas E. *Direct Democracy: The Politics of Initiative, Referendum, and Recall*. Cambridge, Mass.: Harvard University Press, 1989.

Crossen, Cynthia. *Tainted Truth*. New York: Simon & Schuster, 1994.

Crouse, Timothy. *The Boys on the Bus*. New York: Ballantine Books, 1972.

Czitrom, Daniel J. *Media and the American Mind*. Chapel Hill: University of North Carolina Press, 1982.

Dahl, Robert A. *Democracy and Its Critics*. New Haven: Yale University Press, 1989.

Davis, Douglas. *The Five Myths of Television Power.* New York: Simon & Schuster, 1993.

Deakin, James. *Straight Stuff.* New York: William Morrow, 1984.

Dennis, Everette E. *Of Media and People.* Newbury Park, Calif.: Sage Publications, 1992.

Dennis, Everette E., et al. *The Media at War: The Press and the Persian Gulf Conflict.* A Gannett Foundation Report. New York: Gannett Foundation Media Center, Columbia University, 1991.

Devol, Kenneth S., ed. *Mass Media and the Supreme Court.* New York: Hastings House, 1982.

Diamond, Edwin, et al. *Telecommunications in Crisis.* Washington, D.C.: Cato Institute, 1983.

Diggins, John P. *The Lost Soul of American Politics.* Chicago: University of Chicago Press, 1984.

Dill, Barbara, and Martin London. *At What Price?: Libel Law and Freedom of the Press.* New York: The Twentieth Century Fund Press, 1993.

Dionne, E. J., Jr. *Why Americans Hate Politics.* New York: Touchstone Books, 1992.

Disasters and the Mass Media. Washington, D.C.: National Academy of Sciences, 1980.

Donovan, Robert J., and Ray Scherer. *Unsilent Revolution.* Cambridge, England: Cambridge University Press, 1992.

Edsall, Thomas Byrne, and Mary D. Edsall. *Chain Reaction.* New York: W. W. Norton, 1991.

Efron, Edith. *The News Twisters.* Los Angeles: Nash Publishing, 1971.

Elkins, Stanley, and Eric McKitrick. *The Age of Federalism.* New York: Oxford University Press, 1993.

Emery, Walter B. *Broadcasting and Government.* East Lansing: Michigan State University Press, 1961.

Entman, Robert M. *Democracy Without Citizens.* New York: Oxford University Press, 1989.

Epstein, Edward J. *News from Nowhere: Television and the News.* New York: Vintage Books, 1974.

Etzioni, Amitai. *An Immodest Agenda.* New York: McGraw-Hill, 1984.

———. *The Spirit of Community.* New York: Crown Publishers, 1993.

Firestone, Charles M. *The Search for the Holy Paradigm.* Washington, D.C.: The Aspen Institute, 1993.

———, ed. *Television for the 21st Century: The Next Wave.* Washington, D.C.: The Aspen Institute, 1993.

Fish, Stanley. *There's No Such Thing as Free Speech and It's a Good Thing Too.* New York: Oxford University Press, 1994.

Fisher-LaMay, Craig L., ed. *Covering Campaign '88: The Politics of Character and the Character of Politics.* New York: Gannett Center for Media Studies, Columbia University, 1988.

Fishkin, James S. *Democracy and Deliberation: New Directions for Democratic Reform.* New Haven: Yale University Press, 1991.

Forer, Lois G. *A Chilling Effect.* New York: W. W. Norton, 1987.

Frank, Reuven. *Out of Thin Air.* New York: Simon & Schuster, 1991.

Frank, Ronald E., and Marshall G. Greenberg. *The Public's Use of Television.* Newbury Park, Calif.: Sage Publications, 1980.

Friendly, Fred W. *The Good Guys, the Bad Guys and the First Amendment.* New York: Random House, 1975.

———. *Minnesota Rag.* New York: Random House, 1981.

Fukuyama, Francis. *The End of History and the Last Man.* New York: The Free Press, 1992.

Gandy, Oscar H., Jr. *The Panoptic Sort: A Political Economy of Personal Information.* Boulder, Colo.: Westview Press, 1993.

Gans, Herbert J. *Deciding What's News.* New York: Vintage Books, 1980.

———. *Popular Culture and High Culture.* New York: Basic Books, 1974.

Garry, Patrick. *Scrambling for Protection.* Pittsburgh: University of Pittsburgh Press, 1994.

Gasman, Lawrence. *Telecompetition.* Washington, D.C.: Cato Institute, 1994.

Geller, Henry. *Fiber Optics: An Opportunity for a New Policy?* Washington, D.C.: The Annenberg Washington Program in Communications Policy Studies of Northwestern University, 1991.

———. *1995–2005: Regulatory Reform for Principal Electronic Media.* Washington, D.C.: The Annenberg Washington Program in Communications Policy Studies of Northwestern University, 1994.

Germond, Jack W., and Jules Witcover. *Mad as Hell.* New York: Warner Books, 1993.

Gilder, George. *Life After Television.* Knoxville, Tenn.: Whittle Direct Books, 1990.

Gillmor, Donald M. *Power, Publicity and the Abuse of Libel Law.* New York: Oxford University Press, 1992.

Ginsberg, Benjamin, and Martin Shefter. *Politics by Other Means.* New York: Basic Books, 1990.

Gitlin, Todd. *Inside Prime Time.* New York: Pantheon Books, 1983.

Goldenson, Leonard H. *Beating the Odds.* New York: Scribner's, 1991.

Goldman, Peter, and Tony Fuller. *The Quest for the Presidency 1984.* New York: Bantam Books, 1985.

Goldstein, Tom. *The News at Any Cost.* New York: Simon & Schuster, 1985.

Green, Philip, ed. *Democracy, Key Concepts in Critical Theory.* Atlantic Highlands, N.J.: Humanities Press, 1993.

Greider, William. *Who Will Tell the People.* New York: Simon & Schuster, 1992.

Habermas, Jurgen. *The Structural Transformation of the Public Sphere: An Inquiry into a Category of Bourgeois Society,* trans. Thomas Burger. Cambridge, Mass.: MIT Press, 1993.

Halberstam, David. *The Powers That Be.* New York: Knopf, 1979.

Hamilton, Alexander, James Madison, and John Jay. *The Federalist Papers.* New York: Bantam Books, 1982.

Hansen, Phillip. *Hannah Arendt.* Stanford: Stanford University Press, 1993.

Harwood Group. *Citizens and Politics.* Washington, D.C.: Kettering Foundation, 1991.

Heard, Alexander, and Michael Nelson. *Presidential Selection.* Durham: Duke University Press, 1987.

Heim, Michael. *The Metaphysics of Virtual Reality*. New York: Oxford University Press, 1993.

Hemmer, Joseph J. *The Supreme Court and the First Amendment*. New York: Praeger, 1986.

Henry, William A., III. *Visions of America*. Boston: Atlantic Monthly Press, 1985.

Hentoff, Nat. *The First Freedom*. New York: Delacorte, 1988.

Herman, Edward S., and Noam Chomsky. *Manufacturing Consent*. New York: Pantheon Books, 1988.

Hertsgaard, Mark. *On Bended Knee*. New York: Farrar, Straus & Giroux, 1988.

Hill, Melvyn A., ed. *Hannah Arendt: The Recovery of the Public World*. New York: St. Martin's Press, 1979.

Hoyt, Edwin Palmer, Jr. *Jumbos and Jackasses*. New York: Doubleday, 1960.

Iyengar, Shanto, and Donald R. Kinder. *News That Matters*. Chicago: University of Chicago Press, 1987.

Jefferson, Thomas. *The Political Writings of Thomas Jefferson: Representative Selections*, ed. Edward Dumbauld. New York: Liberal Arts Press, 1955.

Kagan, Donald. *Pericles of Athens and the Birth of Democracy*. New York: Touchstone, 1991.

Kahn, Frank J., ed. *Documents of American Broadcasting*. Englewood Cliffs, N.J.: Prentice-Hall, 1978.

Kalb, Marvin. *The Nixon Memo*. Chicago: University of Chicago Press, 1994.

Kalven, Harry, Jr. *A Worthy Tradition*. New York: Harper & Row, 1988.

Karnow, Stanley. *Vietnam: A History*. New York: Viking, 1983.

Key, V. O., Jr. *Politics and Pressure Groups*. New York: Thomas Y. Crowell, 1964.

Klapper, Joseph T. *The Effects of Mass Communication*. New York: The Free Press, 1960.

Kotkin, Joel, and Yoriko Kishimoto. *The Third Century*. New York: Crown Publishers, 1988.

Kubey, Robert, and Minaly Csikszentmihelyi. *Television and the Quality of Life*. Hillsdale, N.J.: Lawrence Erlbaum Associates, 1990.

Lanham, Richard A. *The Electronic Word*. Chicago: University of Chicago Press, 1993.

Lashner, Marilyn A. *The Chilling Effect in TV News*. New York: Praeger, 1987.

Lazarsfeld, Paul F., Bernard Berelson, and Hazel Gaudet. *The People's Choice*. New York: Columbia University Press, 1948.

Lazere, Donald, ed. *American Media and Mass Culture*. Berkeley: University of California Press, 1987.

LeRoy, David J., and Christopher H. Sterling, eds. *Mass News*. Englewood Cliffs, N.J.: Prentice-Hall, 1973.

Levy, Leonard W. *Emergence of a Free Press*. New York: Oxford University Press, 1985.

Lewis, Anthony. *Make No Law*. New York: Random House, 1991.

Lichtenberg, Judith, ed. *Democracy and the Mass Media*. Cambridge, England: Cambridge University Press, 1990.

Lichter, S. Robert, Linda S. Lichter, and Stanley Rothman. *Watching America*. New York: Prentice-Hall Press, 1991.

Lichter, S. Robert, Stanley Rothman, and Linda S. Lichter. *The Media Elite*. Bethesda, Md.: Adler & Adler, 1986.

Lippmann, Walter. *Essays in the Public Philosophy*. Boston: Little, Brown, 1955.

———. *Public Opinion*. New York: The Free Press, 1922.

London, Martin, and Barbara Dill. *At What Price? Libel Law and Freedom of the Press*. New York: Twentieth Century Fund, 1993.

Lorenz, Konrad. *On Aggression*. San Diego: Harcourt Brace Jovanovich, 1966.

Lowi, Theodore J. *The End of Liberalism*. New York: W. W. Norton, 1979.

Macarthur, John R. *Second Front*. New York: Hill & Wang, 1992.

McCullough, David. *Truman*. New York: Simon & Schuster, 1992.

McChesney, Robert W. *Telecommunications, Mass Media and Democracy*. New York: Oxford University Press, 1993.

McDonald, Forrest. *The American Presidency*. Lawrence: University of Kansas Press, 1994.

MacDonald, J. Fred. *One Nation Under Television*. New York: Pantheon Books, 1990.

McKenna, George. *American Populism*. New York: Putnam's Sons, 1974.

McKibben, Bill. *The Age of Missing Information*. New York: Random House, 1992.

McLuhan, Marshall. *The Gutenberg Galaxy*. Toronto: University of Toronto Press, 1962.

———. *Understanding Media,* New York: New American Library, 1964.

MacNeil, Robert. *The People Machine*. New York: Harper & Row, 1968.

Malone, Dumas. *Jefferson, the Sage of Monticello*. Boston: Little, Brown, 1981.

———. *Jefferson and the Rights of Man*. Boston: Little, Brown, 1951.

———. *Jefferson the Virginian*. Boston: Little, Brown, 1948.

Mankiewicz, Frank, and Joel Swerdlow. *Remote Control*. New York: Times Books, 1978.

Mathews, David. *Politics for People*. Urbana: University of Illinois Press, 1994.

Matusow, Barbara. *The Evening Stars*. Boston: Houghton Mifflin, 1983.

Mayer, Martin. *About Television*. New York: Harper & Row, 1972.

———. *Making News*. Boston: Harvard Business School Press, 1993.

Media, Democracy and the Information Highway. New York: Freedom Forum Media Studies Center, Columbia University, 1993.

Media at the Millennium. New York: Freedom Forum Media Studies Center, Columbia University, 1992.

Media Studies Journal. New York: Freedom Forum Media Studies Center, Columbia University, Fall 1988, Spring 1989, Winter 1989, Winter 1990, Fall 1991, Fall 1992, Spring 1994, Winter 1994.

Meier, Christian. *The Greek Discovery of Politics*. Trans. David McLintock. Cambridge, Mass.: Harvard University Press, 1990.

Meyrowitz, Joshua. *No Sense of Place*. New York: Oxford University Press, 1985.

Mickelson, Sig. *From Whistle Stop to Sound Bite*. New York: Praeger, 1989.

———. *The Electric Mirror*. New York: Dodd, Mead, 1972.

Milgram, Stanley, and R. Lance Shotland. *Television and Antisocial Behavior*. New York: Academic Press, 1973.

Minow, Newton N., and Fred H. Cate. *Who Is an Impartial Juror in an Age of*

Mass Media? Washington, D.C.: *American University Law Review,* vol. 40, Winter 1991, No. 2.

Minow, Newton N., and Clifford M. Sloan. *For Great Debates.* New York: Priority Press, 1987.

Monroe, James A. *The Democratic Wish: Popular Participation and the Limits of American Government.* New York: Basic Books, 1992.

Morgan, Edmund S. *Inventing the People.* New York: W. W. Norton, 1988.

Morone, James A. *The Democratic Wish.* New York: Basic Books, 1990.

Morris, Richard B. *Alexander Hamilton and the Founding of the Nation.* New York: Dial Press, 1957.

Naisbitt, John. *Megatrends.* New York: Warner Books, 1982.

Neuman, W. Russell. *The Future of the Mass Audience.* Cambridge, England: Cambridge University Press, 1993.

Neuman, W. Russell, Marion R. Just, and Ann N. Crigler. *Common Knowledge.* Chicago: University of Chicago Press, 1992.

Neustadt, Richard E. *Presidential Power and the Modern Presidents.* New York: The Free Press, 1990.

Nieman Reports. *Can Journalism Shape the New Technologies?* Cambridge, Mass.: Harvard University, 1994.

———. *The War and the Press.* Cambridge, Mass.: Harvard University, 1991.

Nyhan, Michael J., ed. *The Future of Public Broadcasting.* New York: Praeger, 1976.

O'Neil, Michael J. *The Roar of the Crowd.* New York: Times Books, 1993.

Opotowsky, Stan. *TV, the Big Picture.* New York: E. P. Dutton, 1961.

Orren, Gary R., and Nelson W. Polsby, eds. *Media and Momentum: The New Hampshire Primary and Nomination Politics.* Chatham, N.J.: Chatham House, 1987.

Page, Benjamin I., and Robert Y. Shapiro. *The Rational Public: Fifty Years of Trends in Americans' Policy Preferences.* Chicago: University of Chicago Press, 1992.

Paley, William S. *As It Happened.* New York: Doubleday, 1979.

Parenti, Michael. *Inventing Reality.* New York: St. Martin's Press, 1986.

Patterson, Thomas E. *Let the Press Be the Press.* New York: Twentieth Century Fund, 1993.

———. *Out of Order.* New York: Alfred A. Knopf, 1993.

———. *The Mass Media Election.* New York: Praeger, 1980.

Patterson, Thomas E., and Robert D. McClure. *The Unseeing Eye.* New York: Putnam, 1976.

Pease, Otis. *The Responsibilities of American Advertising.* New Haven: Yale University Press, 1958.

Pool, Ithiel de Sola. *Technologies of Freedom.* Cambridge, Mass.: Harvard University Press, 1983.

Postman, Neil. *Amusing Ourselves to Death.* New York: Penguin Books, 1985.

———. *Technopoly.* New York: Alfred A. Knopf, 1992.

Powe, Lucas A., Jr. *American Broadcasting and the First Amendment.* Berkeley: University of California Press, 1987.

———. *The Fourth Estate and the Constitution.* Berkeley: University of California Press, 1991.

Powers, Ron. *The Newscasters*. New York: St. Martin's Press, 1977.

Project on Communications and Information Policy Options. Washington, D.C.: Benton Foundation, 1989.

Randall, Willard Sterne. *Thomas Jefferson: A Life*. New York: HarperCollins, 1993.

Ranney, Austin. *Channels of Power*. New York: Basic Books, 1983.

Rauch, Jonathan. *Demosclerosis*. New York: Times Books, 1994.

Rawls, John. *A Theory of Justice*. Cambridge, Mass.: Harvard University Press, 1971.

———. *Political Liberalism*. New York: Columbia University Press, 1993.

Reichley, A. James. *The Life of the Parties*. New York: The Free Press, 1992.

Reston, James. *The Artillery of the Press*. New York: Harper & Row, 1966, 1967.

Rheingold, Howard. *The Virtual Community: Finding Connection in a Computerized World*. Reading, Mass.: Addison-Wesley, 1993.

Richard, Carl J. *The Founders and the Classics*. Cambridge, Mass.: Harvard University Press, 1994.

Rogers, Lindsay. *The Pollsters*. New York: Alfred A. Knopf, 1949.

Rosenstiel, Tom. *The Beat Goes On*. New York: Twentieth Century Fund, 1994.

———. *Strange Bedfellows: How Television and the Presidential Candidates Changed American Politics, 1992*. New York: Hyperion, 1993.

Ross, Donald K. *A Public Citizen's Action Manual*. New York: Grossman Publishers, 1973.

Rushkoff, Douglas. *Media Virus*. New York: Ballantine Books, 1994.

Sabato, Larry J. *Feeding Frenzy*. New York: The Free Press, 1991.

Schell, Jonathan. *The Real War*. New York: Pantheon Books, 1987.

Schlesinger, Arthur M., Jr. *The Cycles of American History*. Boston: Houghton Mifflin, 1986.

Schorr, Daniel. *The Theodore H. White Lecture*. Cambridge: Joan Shorenstein Barone Center on the Press, Politics and Public Policy, Harvard University, 1993.

Schram, Martin. *The Great American Video Game*. New York: William Morrow, 1987.

Schudson, Michael. *Discovering the News*. New York: Basic Books, 1978.

Selnow, Gary W. *High-Tech Campaigns*. New York: Praeger, 1994.

Selnow, Gary W., and Richard R. Gilbert. *Society's Impact on Television*. New York: Praeger, 1993.

Seventy Hours and Thirty Minutes. New York: NBC News, 1964.

Shalit, Gene, and Lawrence K. Grossman. *Somehow It Works*. Garden City, N.Y.: Doubleday, 1965.

Simmons, Steven J. *The Fairness Doctrine and the Media*. Berkeley: University of California Press, 1978.

Smolla, Rodney A. *Free Speech in an Open Society*. New York: Alfred A. Knopf, 1992.

Smoller, Fredric T. *The Six O'Clock Presidency*. New York: Praeger, 1990.

Spender, Stephen. *World Within World*. New York: St. Martin's Press, 1994.

Sperber, A. M. *Murrow: His Life and Times*. New York: Freundlich Books, 1986.

Stacks, Don W., ed. *Seminar on Direct Electronic Democracy*. Miami: University of Miami School of Communications, 1993.

Steiner, Gary A. *The People Look at Television.* New York: Alfred A. Knopf, 1963.

Stoler, Peter. *The War Against the Press.* New York: Dodd, Mead, 1986.

Sunstein, Cass R. *Democracy and the Problem of the First Amendment.* New York: The Free Press, 1993.

Sussman, Barry. *What Americans Really Think.* New York: Pantheon Books, 1988.

Swerdlow, Joel L., ed. *Media Technology and the Vote.* Washington, D.C.: The Annenberg Washington Program, 1988.

Taylor, Paul. *See How They Run.* New York: Alfred A. Knopf, 1990.

The Debate on the Constitution. New York: The Library of America, 1993.

The Homestretch, New Politics. New Media. New Voters? New York: The Freedom Forum Media Studies Center, 1992.

The Media and Campaign '92. New York: Freedom Forum Media Studies Center, 1993.

The Twentieth Century Fund Task Force on Public Television. *Quality Time?* New York: The Twentieth Century Fund Press, 1993.

Tinker, Grant, and Bud Rukeyser. *Tinker in Television.* New York: Simon & Schuster, 1994.

Toffler, Alvin. *The Third Wave.* New York: Bantam Books, 1980.

Tocqueville, Alexis de. *Democracy in America,* vols. 1 and 2. New York: Alfred A. Knopf, 1945.

Weaver, Paul H. *News and the Culture of Lying.* New York: The Free Press, 1994.

Westbrook, Robert B. *John Dewey and American Democracy.* Ithaca: Cornell University Press, 1991.

White, Theodore H. *The Making of the President 1960, 1964, 1968, 1972.* New York: Atheneum, 1961, 1965, 1969, 1973.

Whittemore, Hank. *CNN: The Inside Story.* Boston: Little, Brown, 1990.

Will, George F. *Restoration.* New York: The Free Press, 1992.

———. *The Leveling Wind.* New York: Viking, 1994.

William Benton Conference. *Campaigning on Cue.* Chicago: University of Chicago Press, 1988.

Williams, Frederick, and John V. Pavlik, eds. *The People's Right to Know.* Hillsdale, N.J.: Lawrence Erlbaum Associates, 1994.

Williams, Huntington. *Beyond Control: ABC and the Fate of the Networks.* New York: Atheneum, 1989.

Wills, Gary. *Certain Trumpets.* New York: Simon & Schuster, 1994.

Winn, Marie. *The Plug-in Drug.* New York: Viking, 1977.

Wood, Gordon S. *The Creation of the American Republic.* New York: W. W. Norton, 1969.

———. *The Radicalism of the American Revolution.* New York: Vintage Books 1993; originally published by Alfred A. Knopf, 1991.

Wriston, Walter B. *The Twilight of Sovereignty.* New York: Charles Scribner's Sons, 1992.

Yankelovich, Daniel. *Coming to Public Judgment.* Syracuse: Syracuse University Press, 1991.

INDEX